Collins

Secure Maths
Year 6

a primary maths intervention programme

Pupil Resource Pack

Collins

William Collins' dream of knowledge for all began with the publication of his first book in 1819.
A self-educated mill worker, he not only enriched millions of lives, but also founded a flourishing publishing house. Today, staying true to this spirit, Collins books are packed with inspiration, innovation and practical expertise. They place you at the centre of a world of possibility and give you exactly what you need to explore it.

Collins. Freedom to teach.

An imprint of HarperCollins*Publishers*
The News Building
1 London Bridge Street
London
SE1 9GF

Browse the complete Collins catalogue at
www.collins.co.uk

10 9 8 7 6 5 4 3 2 1

ISBN 978-00-0-822152-2

British Library Cataloguing in Publication Data
A catalogue record for this publication is available from the British Library.

Author Bobbie Johns
Publishing manager Fiona McGlade
Development editor Fiona Tomlinson
Editor Nina Smith
Project managed by Alissa McWhinnie, QBS Learning
Copyedited by Catherine Dakin
Proofread by Cassie Fox
Answers checked by Deborah Dobson
Cover design by Amparo Barrera and ink-tank and associates
Cover artwork by Amparo Barrera
Internal design by 2Hoots publishing services
Typesetting by QBS Learning
Illustrations by QBS Learning
Production by Rachel Weaver
Printed and bound by CPI

Contents

Unit 1: Read, write, order and compare numbers up to 10 000 000 and determine the value of each digit

1. Write these numbers in numerals.

a) three hundred and forty thousand, five hundred and six

b) eight hundred thousand and four

c) five million, nine hundred and sixty-four thousand, seven hundred and twenty

d) two million, five thousand and thirty

2. Write these numbers in words.

a) 609 013 ______

b) 8 091 105 ______

3. Calculate, using place value.

a) 400 000 + 500 =

b) 50 000 + 70 =

c) 300 000 + 50 000 + 7000 + 800 + 20 + 1 =

d) 6 000 000 + 700 000 + 40 000 + 1000 + 600 + 50 + 9 =

e) 80 000 + 6 + 4 000 000 + 5000 + 300 =

4. What is the value of the **6** in these numbers?

a) 4 369 215

b) 7 615 004

c) 8 426 119

d) 6 927 354

5. Write these numbers in order, from largest to smallest.

a) 978 935 1 245 742 99 999 987 276

______

b) 5 475 312 5 457 321 5 475 321 5 457 231

______

Unit 1: Read, write, order and compare numbers up to 10 000 000 and determine the value of each digit

1. Write these numbers in numerals in the place-value grid.

a) five hundred and seventy-six thousand and fourteen

b) three million, eight hundred and ninety thousand, four hundred and five

c) three hundred and eighty thousand, one hundred and twelve

d) four million, five hundred and twenty thousand, two hundred and ten

e) four hundred and thirty thousand and ninety-one

f) five million and forty-five

Million	Hundred thousand	Ten thousand	Thousand	Hundred	Ten	Ones

2. What is the value of the **9** in these numbers?

a) 3 196 385 [ ]

b) 8 904 215 [ ]

3. Write these numbers in order, from smallest to largest.

2 150 001 2 090 003 2 125 786 2 089 998

[ ] [ ] [ ] [ ]

Unit 2: Round any whole number to a required degree of accuracy

1. Round 325 816 to the nearest:

a) 10

b) 100

c) 1000

d) 10 000

e) 100 000

2. Draw a line to connect each decimal to the whole number it rounds to. Cross out the decimals that do not connect to either number.

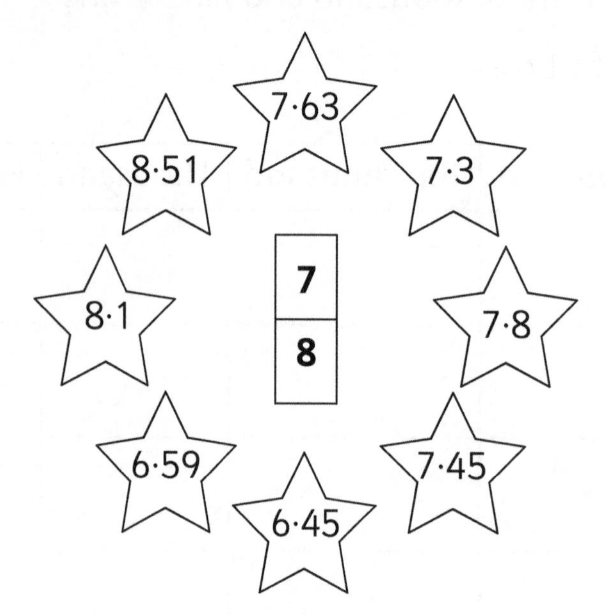

3. Complete the table.

Population	Million
4 000 000	4 million
2 000 000	
	6·5 million
8 400 000	
	5·41 million
7 260 000	

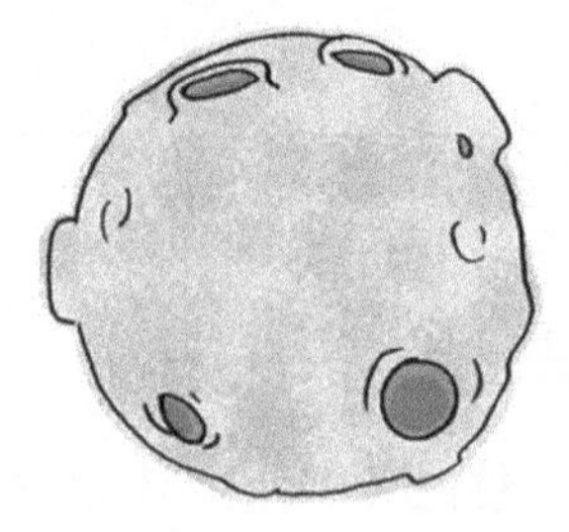

4. The moon is an average of 384 400 kilometres away from earth. What is this distance, to the nearest 1000 kilometres?

Unit 2: Round any whole number to a required degree of accuracy

1. Round 582 971 to the nearest:

a) 100 000

b) 100

2. The area of Australia is 7·69 million square kilometres.
Write this as a seven-digit number.

3. 386 097 is rounded to the nearest 10 and the nearest 100. What do you notice? Is one more accurate than the other?

4. Round 48·36 to the nearest:

a) tenth

b) whole number

c) ten

Unit 3: Use negative numbers in context, and calculate intervals across 0

1. The table shows the average daily January temperatures of some European cities.

City	Temperature °C
Helsinki	–2
Barcelona	13
Moscow	–8
Kiev	–3
Copenhagen	3

a) Which city has the lowest temperature? ____________

b) How much colder is it in Kiev than in Barcelona? [ ] °C

c) What is the difference in temperature between Helsinki and Moscow?

[ ] °C

d) One year, the temperature in Copenhagen was 5 °C lower than the average. What was the temperature that year? [ ] °C

2. Use the symbols > or < to connect these pairs of numbers.

a) –5 [ ] 2 **b)** –1 [ ] –7 **c)** 0 [ ] –1

3. Answer these questions.

a) What is 6 less than 4? [ ]

b) What is 8 less than 5? [ ]

c) What is 5 less than –3? [ ]

d) What is 5 more than –2? [ ]

e) What is 4 more than –6? [ ]

Unit 3: Use negative numbers in context, and calculate intervals across 0

1. The temperature drops from 11°C to –4°C. By how many degrees has the temperature dropped?

☐ °C

2. How much colder is –15°C than –5°C?

☐ °C

3. Order these numbers, from smallest to largest.

9	–4	0	–3	1	–10

☐ ☐ ☐ ☐ ☐ ☐

4. What number do you get when you count back 7 from 5?

☐

5. One cold day, the temperature in Toronto was –3°C. It was 5° colder in St Petersburg. What was the temperature in St Petersburg?

☐ °C

Unit 4: Solve number and practical problems involving number and place value

1. The table shows the land areas of some countries in South America.

Country	Land area in km²	To nearest 100 km²	Millions to two decimal places
Bolivia	1 083 301	1 083 000	1·08 million
Peru	1 279 996		
Mexico	1 943 945		
Venezuela	882 050		
Colombia	1 038 700		
Brazil	8 460 415		

a) Complete the third column by rounding each area to the nearest **100** square kilometres.

b) Complete the fourth column by writing the number as a decimal of a million, to **two decimal places**.

c) Which country has more land, Bolivia or Colombia? ____________

d) Write the countries in order of land size, from smallest to largest.

____

e) Write the land area of Venezuela in words.

____

2. Each child is holding one of these number cards. Work out who has each number.

Ahmed: 'My number has seventy thousands.' ____

Finn: 'My number rounds to 70 600, to the nearest hundred.' ____

Gita: 'My number has five tens.' ____

Saniska: 'My number rounds to 70 000, to the nearest thousand.' ____

Unit 4: Solve number and practical problems involving number and place value

1. The area of Lake Huron is 59 596 square kilometres. What is its area, to the nearest thousand square kilometres?

[ ] km^2

2. Lake Malawi has an area of 30 044 square kilometres. Write the area of Lake Malawi in words.

____

3. The distance between Cairns and Perth, in Australia, is 5954 km.

What is the distance:

a) to the nearest 10 km? [ ]

b) to the nearest 100 km? [ ]

4. Use the clues to work out the mystery number.

- It has less than 6 ones.
- To the nearest ten, it rounds **up** to 41 100.
- The mystery number is [ ].

5. Circle two numbers that add together to make 1 million.

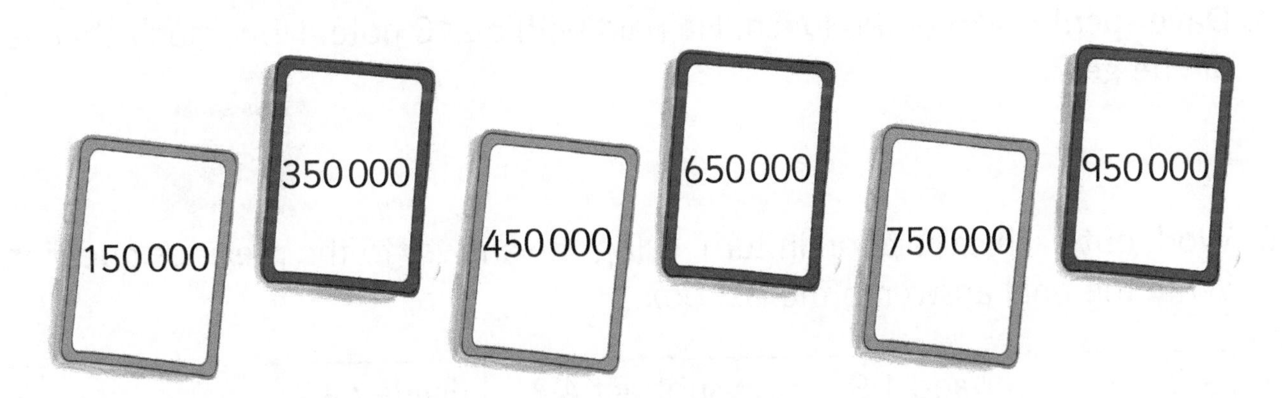

Unit 5: Perform mental calculations, including with mixed operations and large numbers

1. Draw a line to join each calculation to a suitable strategy.

Calculation	Strategy
3765 + 199	Known multiplication fact
0·8 × 9	Near multiple of 10 and adjust
1000 – ☐ = 650	Place-value subtraction
2735 + 28	Near multiple of 100 and adjust
3857 – 200	Number bond to 100

2. Calculate mentally.

a) 3746 – 300 = ☐

b) 1·2 × 3 = ☐

c) 6·8 – 2·6 = ☐

d) 299 + 194 = ☐

e) 1·8 + 4·5 = ☐

f) 5 + 0·3 + 0·04 + 0·002 = ☐

3. Calculate the value of the symbol Δ in each calculation.

a) 630 + Δ = 1000 Δ = ☐

b) 3600 ÷ Δ = 90 Δ = ☐

c) 9283 + Δ = 9583 Δ = ☐

d) 0·6 × Δ = 4·2 Δ = ☐

4. Dave spent £6·65 on his lunch. He paid with a £10 note. How much change did he get?

£ ☐

5. Work out each calculation in turn, using the answer to the previous question. Write the final answer in the last box.

3·2 × 3	add 1·9	subtract 4·3	divide by 2	

Unit 5: Perform mental calculations, including with mixed operations and large numbers

1. Explain how to add 99 to 3158.

____

____

2. Calculate.

a) 2867 + 100 = ☐

b) 275·6 × 0 = ☐

c) 0·9 × 2 = ☐

d) 10 – 3·8 = ☐

3. Write the missing number to each calculation in the box.

a) 1000 – ☐ = 120

b) 1·2 ÷ ☐ = 3

c) 25·48 × ☐ = 25·48

d) 2 + 0·30 + ☐ = 2·37

4. 34 + 66 = 100

a) Write two related larger number calculations: one addition and one subtraction.

____

b) Write two related decimal calculations: one addition and one subtraction.

____

Unit 6: Use estimation to check answers to calculations and determine, in the context of a problem, an appropriate degree of accuracy

1. Estimate the numbers shown on the number line.

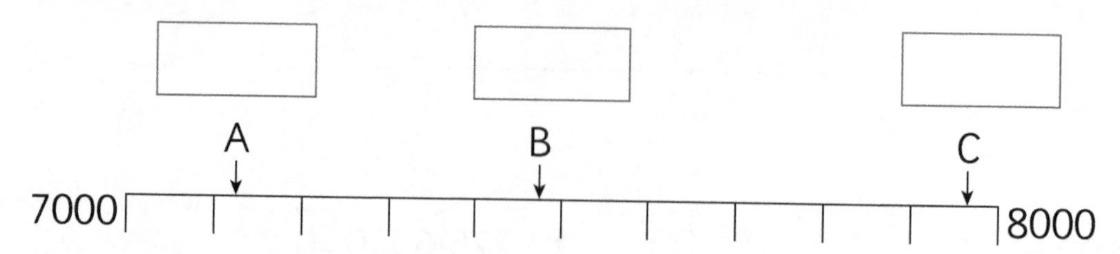

2. Calculate an approximate answer, using rounding. Write the rounded numbers you have used.

a) 9365 – 3856 ≈ ☐ – ☐ ☐ = ☐

b) 48 + 2856 + 319 ≈ ☐ + ☐ + ☐ = ☐

c) 4·95 × 3·6 ≈ ☐ × ☐ = ☐

d) 36·96 ÷ 8·8 ≈ ☐ ÷ ☐ = ☐

e) 12 096 + 11 782 ≈ ☐

3. The digits are correct, but all the answers are wrong. Use rounding, or other strategies, to find the correct answers.

a) 12·7 × 10·05 = 1276·35 ☐

b) 119 ÷ 12·5 = 95·2 ☐

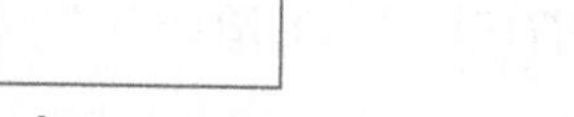

4. Here are population figures for the countries of the United Kingdom.

Country	Population
England	53 012 500
Scotland	5 295 000
Wales	3 063 400
N Ireland	1 810 900

a) What is the approximate total population, **in millions**, of the UK?

☐

b) Approximately how many more people live in Wales than in N Ireland?

☐

Unit 6: Use estimation to check answers to calculations and determine, in the context of a problem, an appropriate degree of accuracy

1. Write the letter shown next to each number in the correct place on the number line. **a)** has been done for you.

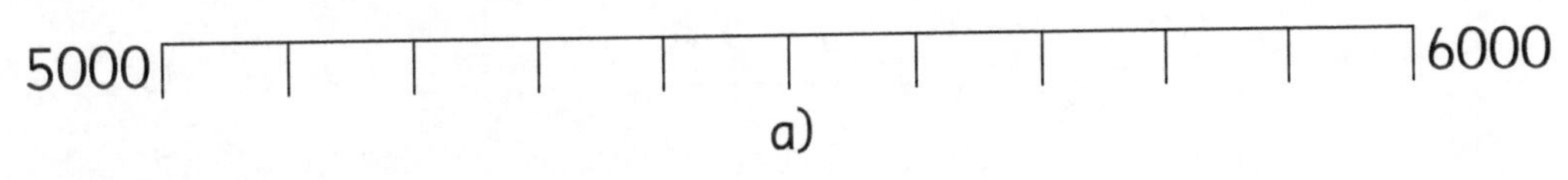

a) 5510 **b)** 5005

c) 5875 **d)** 5325

2. Calculate an approximate answer to these calculations. Show your working.

a) 12 945 – 9782 ≈

b) 10·32 × 9·75 ≈

c) 99·06 ÷ 25·4 ≈

3. Show that 23·8 × 3·99 = 104·92 cannot be correct.

4. Calculate an approximate total for this list of items.

Total:

Unit 7: Multiply multi-digit numbers up to four digits by a two-digit whole number using the formal written method of long multiplication

1. Calculate the answers using your chosen column method.

a)
```
  4 3 1 2
×     1 7
```

b)
```
  3 0 9 4
×     2 6
```

2. Calculate the answers using your chosen column method.

a) 3084 × 32

b) 1947 × 43

3. A farmer supplies eggs in trays of 36.
This year, the farmer supplied 1275 trays.
How many eggs is that in total?

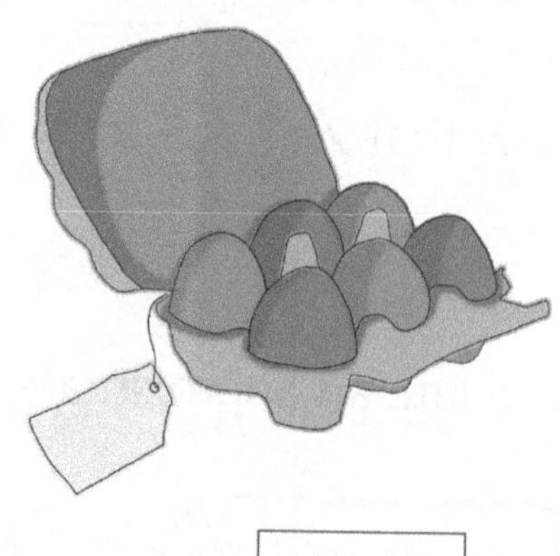

☐ eggs

4. On a sheet of A4-squared paper, there are 28 rows of 18 squares in each row.
How many squares are on the sheet?

☐ squares

Unit 7: Multiply multi-digit numbers up to four digits by a two-digit whole number using the formal written method of long multiplication

1. Calculate 2068 × 14 using a column method.

2. Calculate 2756 × 45 using a column method.

3. The average mass of the car a factory produces is 1985 kg. What is the total mass of 54 of the cars?

kg

Unit 8: Divide numbers up to four digits by a two-digit whole number using the formal written method of long division and interpret remainders as whole number remainders, fractions or by rounding, as appropriate for the context

1. Write the first five multiples of 16.

2. Use the multiples of 16 from Question 1, and your preferred subtraction method of division, to calculate:

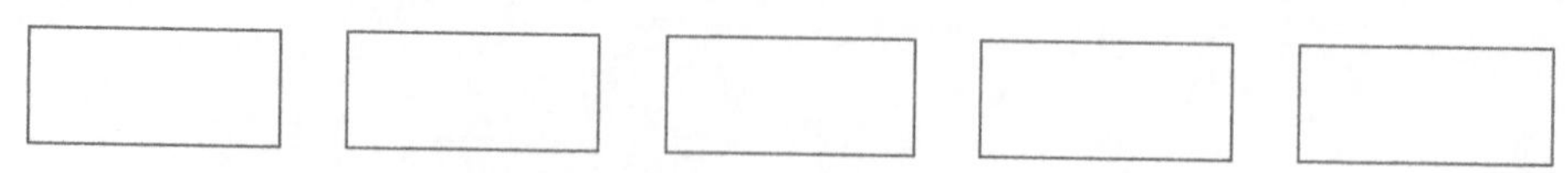

a) $16\overline{)8736}$

b) $16\overline{)6176}$

3. Calculate 6624 ÷ 24.

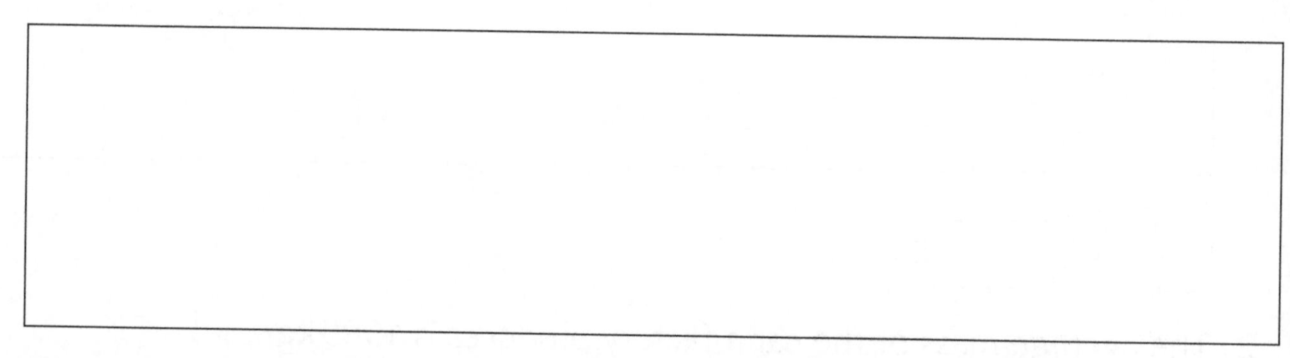

4. Marbles are packed in packets of 36. How many packets are needed for 8856 marbles?

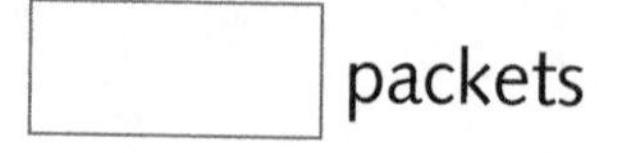

packets

5. Calculate these divisions, writing each remainder as a fraction.

a) 3867 ÷ 5 =

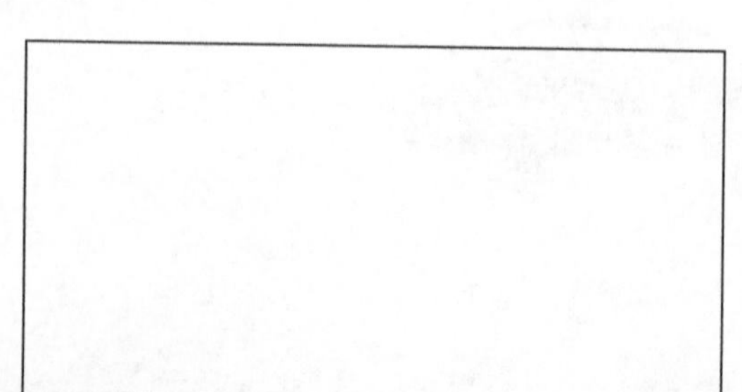

b) 2603 ÷ 8 =

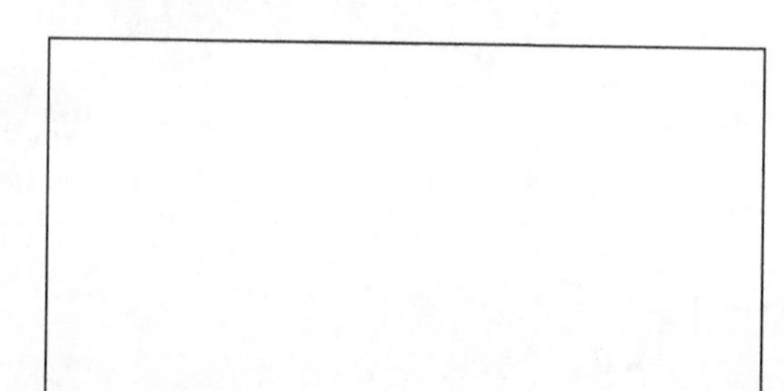

Unit 8: Divide numbers up to four digits by a two-digit whole number using the formal written method of long division and interpret remainders as whole number remainders, fractions or by rounding, as appropriate for the context

1. Calculate 8820 ÷ 12.

2. Eggs are packed in trays of 18. How many trays are needed for 8694 eggs?

3. Work out 8556 ÷ 23.

4. Write a number between 70 and 90 that will have a remainder of 2 when it is divided by 6.

5. Calculate 7365 ÷ 12 giving the remainder as a fraction.

6. Dave has 375 m of cable. He cuts it into 21 equal lengths. How long is each piece? Give any remainder as a fraction.

Unit 9: Divide numbers up to four digits by a two-digit number using the formal written method of short division where appropriate, interpreting remainders according to the context

1. Calculate these divisions, using the short division method.

a) 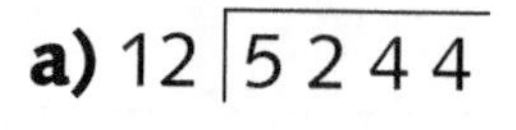

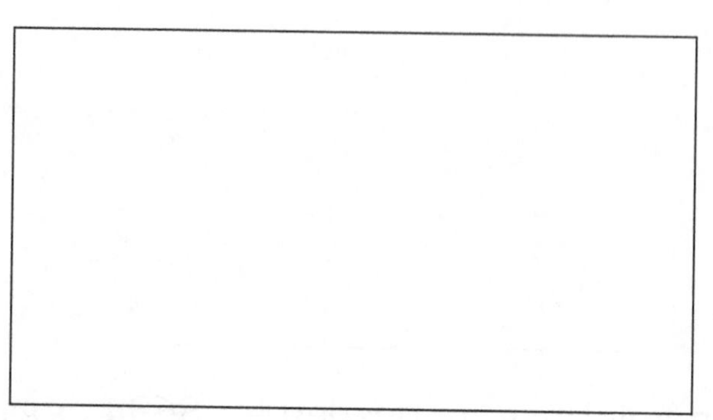

b) 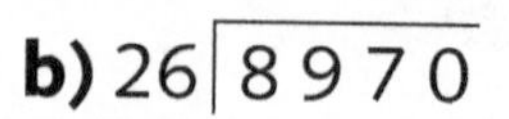

2. Calculate these.

a) 5112 ÷ 18

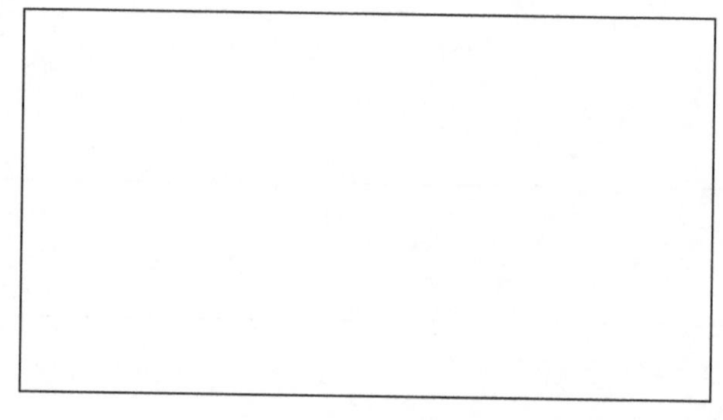

b) 4992 ÷ 32

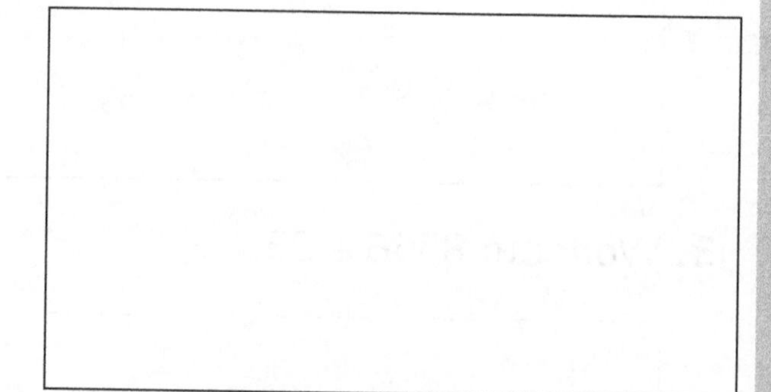

3. Calculate 3756 ÷ 16, giving the remainder as a simplified fraction.

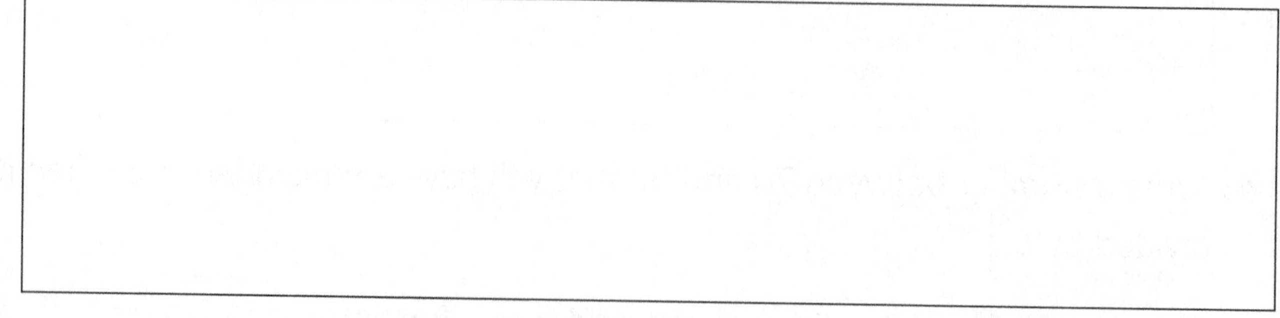

4. Pencils are packed in sets of 24. How many packets are needed for 4700 pencils?

☐ packets

Unit 9: Divide numbers up to four digits by a two-digit number using the formal written method of short division where appropriate, interpreting remainders according to the context

1. Calculate 7476 ÷ 14, using the short division method.

2. Calculate 8602 ÷ 23, using the short division method.

3. Calculate 4646 ÷ 24, giving the remainder as a fraction.

4. Millie uses 36 beads to make a necklace. She has 8960 beads. How many necklaces can she make?

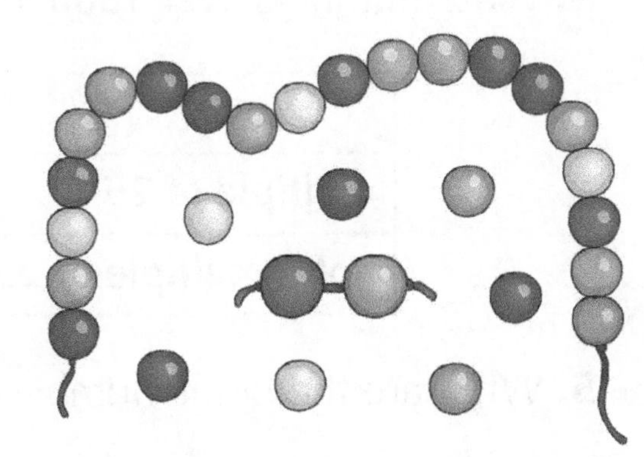

 necklaces

Unit 10: Identify common factors, common multiples and prime numbers

1. Choose **all** the numbers in the box that match the description. You can use each number more than once.

1	2	6	9	11	15	18	20	29	35	36

a) Prime numbers ______________________

b) Multiples of 3 ______________________

c) Common multiples of 6 and 9 ______________________

d) Factors of 36 ______________________

e) Common factors of 24 and 30 ______________________

2. Rearrange these digit cards to make:

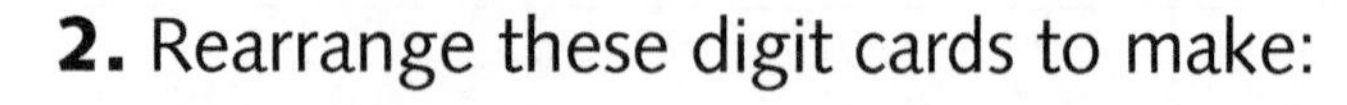

3 4 2

a) a multiple of 4 [ ]

b) a multiple of 6 [ ]

c) the number closest to a multiple of 25 [ ]

3. Put a tick under the numbers which are factors of 2340 and 1275. Put a cross if they are not factors.

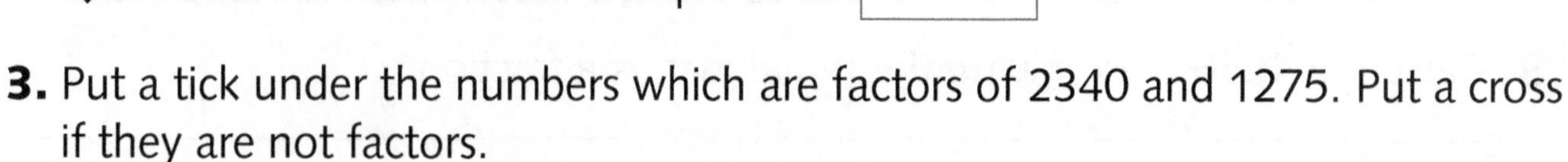

	2	3	4	5	6	9	10	25
2340	√							
1275	x							

4. Write numbers **over 1000** in each part of the Carroll diagram.

	odd	**even**
multiple of 25		
not a multiple of 25		

5. What are the prime numbers between 60 and 70? [ ]

Unit 10: Identify common factors, common multiples and prime numbers

1. Write each number in the correct place on the Venn diagram.

2   4   8   10   15   20   40

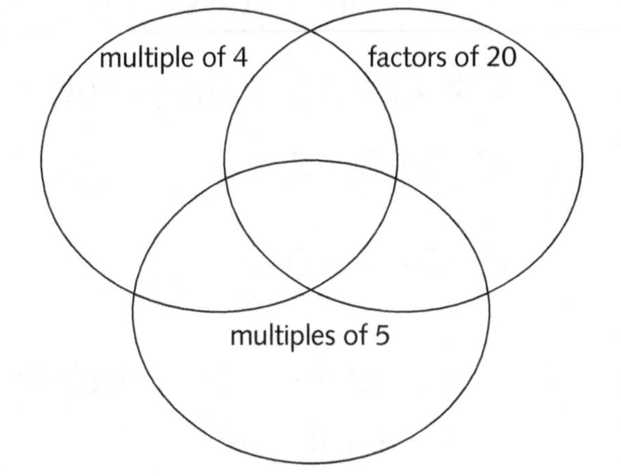

2. Find a common multiple of 5 and 9 which is **over 50**.

3. Answer these.

a) Write **all** the factors of 24.

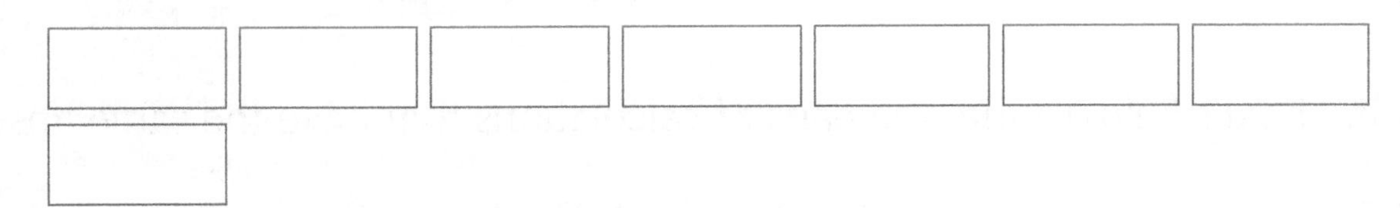

b) Write **all** the factors of 36.

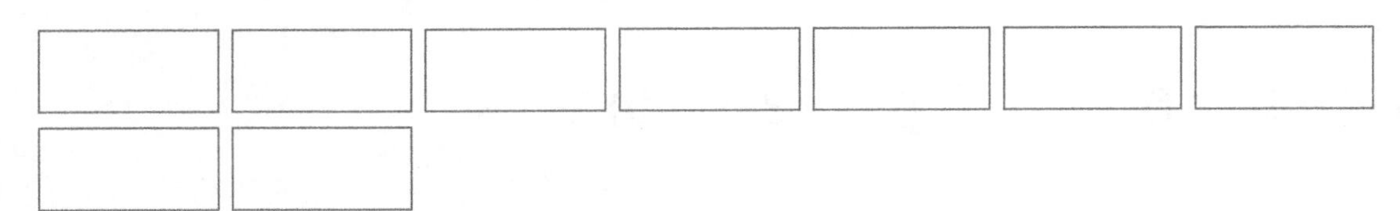

c) Write the **common factors** of 24 and 36.

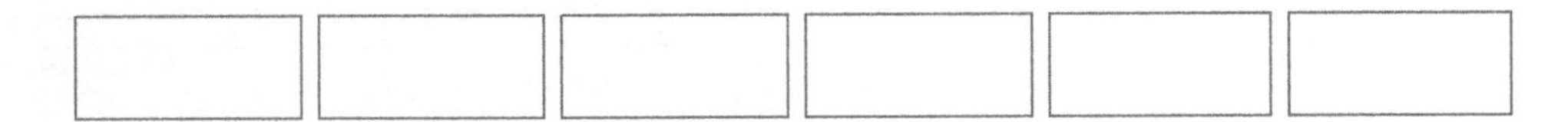

4. Choose a number from the grid that is true for each statement.

70	71	72	73	74
75	76	77	78	79
80	81	82	83	84
85	86	87	88	89

a) 5 less than a multiple of 12

b) a common multiple of 6 and 9

Unit 11: Use their knowledge of the order of operations to carry out calculations involving the four operations

1. Calculate, using the priority of operations.

a) $15 - 3 \times 2 =$ ☐ **b)** $24 - 12 \div 4 =$ ☐

c) $10 - 2 \times 0 =$ ☐ **d)** $5 \times 6 \div 2 =$ ☐

2. Add brackets to the calculations to make the answers correct.

a) $3 \times 4 + 6 = 30$ **b)** $12 - 2 \times 5 = 50$

3. Calculate.

a) $(3 \times 5) + (32 \div 8) =$ ☐ **b)** $(42 \div 6) - (25 \div 5) =$ ☐

4. In the boxes below, write the pairs of calculations that have the same answer.

a) $5^2 - 4$ **b)** $3^2 - 2$ **c)** $3^2 \times 5$ **d)** $4^2 + 5$

e) $9^2 - 6$ **f)** $6^2 + 9$ **g)** $5^2 \times 3$ **h)** $2^2 + 3$

$5^2 - 4$ = ☐ ☐ = ☐

☐ = ☐ ☐ = ☐

5. What is the difference between $5^2 + 3^2$ and $(5 + 3)^2$? ☐

6. Calculate.

a) $7^2 - 5 \times 6 =$ ☐ **b)** $8^2 - (24 \div 6) + 3 \times 10 =$ ☐

Unit 11: Use their knowledge of the order of operations to carry out calculations involving the four operations

1. Circle the correct answer for each calculation.

a) $20 - 5 \times 3 =$ 45 5

b) $16 + 4 \div 2 =$ 10 18

2. Calculate these.

a) $80 - 56 \div 8 =$ ☐ **b)** $10^2 - 10 =$ ☐

3. Write the correct operations in these calculations.

a) 30 ☐ 4 ☐ 5 = 10

b) 15 ☐ 3 ☐ 4 = 20

4. Calculate the difference between $31 + 9 \times 7$ and $(31 + 9) \times 7$.

5. Andi says the answer to $6^2 - 4^2$ is the same as $(6 - 4)^2$. Is Andi right? Show how you know.

Unit 12: Solve addition and subtraction multi-step problems in contexts, deciding which operations and methods to use and why

1. Two numbers have a difference of 4·96. One number is 15·8. Find the other two numbers.

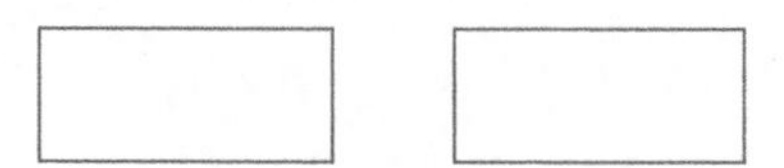

2. What number is halfway between 2·4 and 6·7?

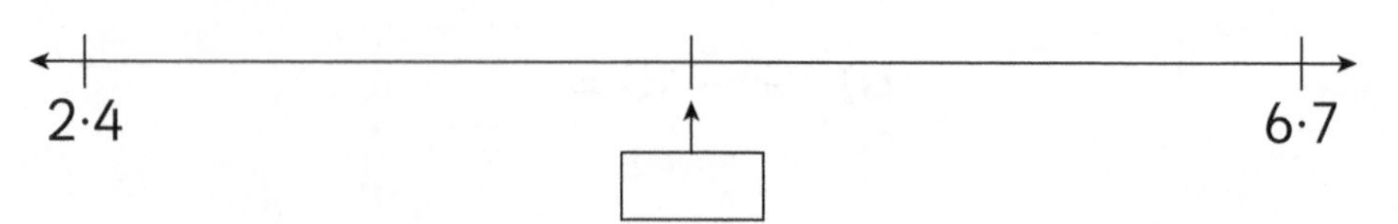

3. Work out the missing digits in this addition.

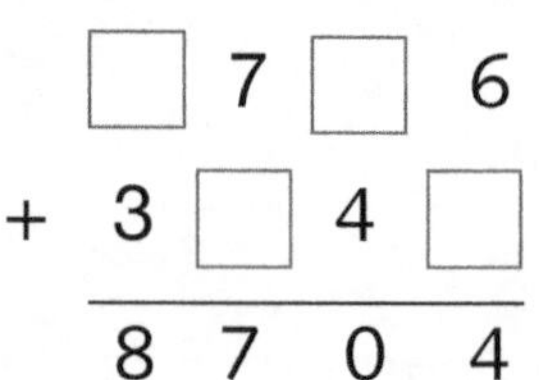

4. Here is a price list for some cakes. Ruben buys one of each size and pays with a £20 note. What change does he get?

£ □

Cakes

Size	Prices
Large	£7.95
Medium	£5.50
Small	£4.25

5. The table shows the amounts of money a company has given to various charities. The company's aim is to give away £1 000 000. How much more does it need to give away to reach this target?

A	£ 91 450
B	£375 195
C	£515 080
D	£

Unit 12: Solve addition and subtraction multi-step problems in contexts, deciding which operations and methods to use and why

1. Which of the numbers below is:

a) closest to 600 ☐

b) furthest from 600 ☐

699	581	621	701	498

2. Work out the missing digits in the subtraction.

```
   7 ☐ 4 ☐
 – ☐ 9 ☐ 5
 ---------
   3 2 6 8
```

3. Three numbers add up to 10. One number is 3·76. What could the other two numbers be?

3·76 + ☐ + ☐ = 10

4. Kai is running a marathon of 42 kilometres. In the first hour, he completed 15·44 kilometres. In the second hour, he ran 14·368 kilometres. How far has he still to run?

☐ kilometres

5. These are the estimated populations of three cities in 2014.

City	Population
Leeds	766 399
Manchester	514 417
Birmingham	1 101 360

a) What is the total population of these three cities? ☐

b) How many more people live in Leeds and Manchester **combined** than in Birmingham?

☐ people

Unit 13: Solve problems involving addition, subtraction, multiplication and division

1.

Country	Square kilometres
England	130 395
Scotland	78 772
Wales	20 779
N Ireland	13 843

a) How much smaller is the combined area of Scotland, Wales and N Ireland than the area of England?

☐ square kilometres

b) The area of the state of Texas, USA, is about 33 times the area of Wales. What is the approximate area of Texas?

☐ square kilometres

c) The area of N Ireland is approximate 14 times the area of Luxembourg. What is the approximate area of Luxembourg?

☐ square kilometres

2. Tarun has read $\frac{7}{10}$ of his book. The book has 340 pages. How many pages has he read?

☐ pages

3. A bakery makes 3640 bread rolls each week. How many rolls does it make in a year (52 weeks)?

☐ rolls

4. In the piece of music a band is playing, the drummer hits the bass drum every 3 seconds and the snare drum, every 6 seconds. How many times in **1 minute** will he play both drums together?

☐ times

Unit 13: Solve problems involving addition, subtraction, multiplication and division

1. Luke has 12 orange flowers, 18 red flowers and 24 white flowers. He wants to make bunches of flowers of one colour with the same number of flowers in each bunch. How many of each colour could be in a bunch of flowers?

☐ or ☐ flowers in each bunch.

2. In 10 years, the population of a village has increased 14 times to 9450 people. How many people lived in the village 10 years ago?

☐ people

3. The table shows one week's sales of the most popular flavour of crisps at a store.

Flavour	Number
ready salted	23 985
salt and vinegar	140 786
cheese and onion	99 005

a) What is the total number of these flavours of crisps the store sold?

☐

b) How many more salt and vinegar crisps were sold than ready salted?

☐

4. A factory packs 25 marbles into each packet. It has made 6245 marbles. How many packets will be needed?

☐ packets

5. 3756 ÷ 16 = 234 remainder 12. Write the answer with a fraction remainder.

☐

Unit 14: Use common factors to simplify fractions; use common multiples to express fractions in the same denomination

1. Write the division number to show which common factor has been used to simplify each fraction.

a)

÷ ☐

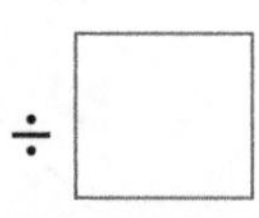

$\frac{15}{25} = \frac{3}{5}$

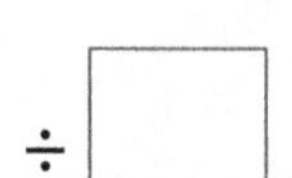

÷ ☐

b)

÷ ☐

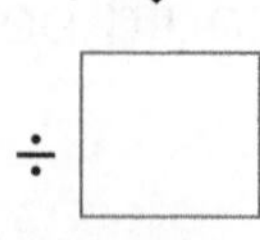

$\frac{20}{24} = \frac{5}{6}$

÷ ☐

c)

÷ ☐

$\frac{24}{30} = \frac{4}{5}$

÷ ☐

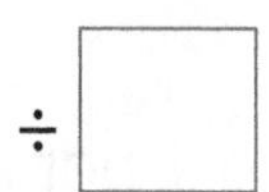

2. Draw a line to match each fraction with its simplified form.

$\frac{6}{12}$ $\frac{9}{12}$ $\frac{8}{10}$ $\frac{12}{20}$ $\frac{70}{100}$

$\frac{3}{4}$ $\frac{3}{5}$ $\frac{7}{10}$ $\frac{1}{2}$ $\frac{4}{5}$

3. Simplify these fractions fully.

a) $\frac{35}{50} = \frac{\square}{\square}$ **b)** $\frac{80}{100} = \frac{\square}{\square}$ **c)** $\frac{32}{4} = \frac{\square}{\square}$

Unit 14: Use common factors to simplify fractions; use common multiples to express fractions in the same denomination

4. Change each pair of fractions into fractions with the same denominator.

a) $\frac{3}{5}$ and $\frac{3}{4}$ $\qquad \frac{3}{5} = \frac{\square}{\square} \qquad \frac{3}{4} = \frac{\square}{\square}$

b) $\frac{1}{4}$ and $\frac{5}{6}$ $\qquad \frac{1}{4} = \frac{\square}{\square} \qquad \frac{5}{6} = \frac{\square}{\square}$

c) $\frac{7}{10}$ and $\frac{2}{3}$ $\qquad \frac{7}{10} = \frac{\square}{\square} \qquad \frac{2}{3} = \frac{\square}{\square}$

5. Which is larger, $\frac{5}{12}$ or $\frac{4}{9}$? Show how you know using equivalent fractions.

Unit 14: Use common factors to simplify fractions; use common multiples to express fractions in the same denomination

1. Simplify these fractions fully.

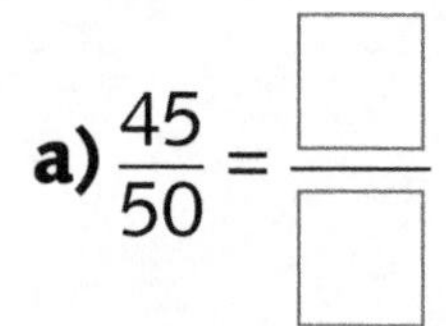

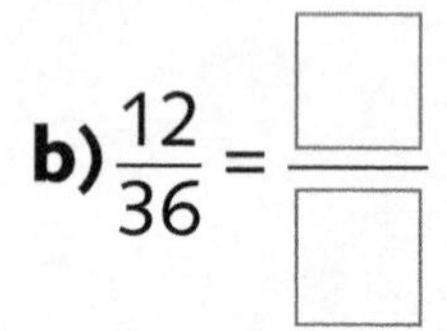

2. Write the missing digits in these equivalent fractions.

$$\frac{\square}{3} = \frac{9}{27} = \frac{3}{\square}$$

3. Change $\frac{3}{10}$ and $\frac{3}{4}$ into equivalent fractions with the same denominator.

$\frac{3}{10} = \frac{\square}{\square}$ $\qquad$ $\frac{3}{4} = \frac{\square}{\square}$

Unit 14: Use common factors to simplify fractions; use common multiples to express fractions in the same denomination

4. Show which is smaller, $\frac{7}{9}$ or $\frac{4}{5}$, using equivalent fractions with the same denominator.

$\frac{7}{9} = \frac{\square}{\square}$ $\frac{4}{5} = \frac{\square}{\square}$

$\frac{\square}{\square}$ is smaller.

Unit 15: Compare and order fractions, including fractions > 1

1. Write one fraction in each part of the Carroll diagram.

$\frac{5}{8}$ $\frac{1}{3}$ $\frac{5}{6}$ $\frac{3}{5}$ $\frac{1}{6}$ $\frac{3}{8}$ $\frac{1}{5}$ $\frac{4}{5}$

	odd denominator	**even denominator**
$< \frac{1}{4}$		
between $\frac{1}{4}$ **and** $\frac{1}{2}$		
between $\frac{1}{2}$ **and** $\frac{3}{4}$		
$> \frac{3}{4}$		

2. Use the symbols $<$ and $>$ to compare these pairs of fractions.

a) $\frac{4}{5}$ ☐ $\frac{7}{8}$ **b)** $\frac{3}{7}$ ☐ $\frac{5}{9}$

c) $\frac{2}{3}$ ☐ $\frac{7}{12}$ **d)** $\frac{5}{6}$ ☐ $\frac{8}{9}$

3. Circle the larger of the two numbers in each pair.

a) $\frac{3}{10}$ $\frac{1}{3}$ **b)** $\frac{4}{7}$ $\frac{5}{8}$ **c)** $1\frac{3}{5}$ $1\frac{5}{9}$

4. Prove that $2\frac{3}{4}$ is more than $\frac{12}{5}$.

5. Order these fractions, from smallest to largest.

a) $\frac{19}{30}$ $\frac{9}{10}$ $\frac{3}{5}$ $\frac{11}{15}$

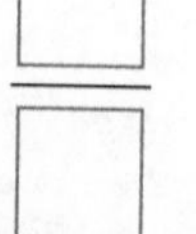
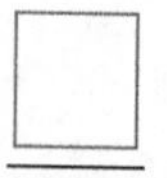
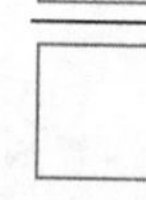
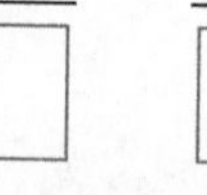
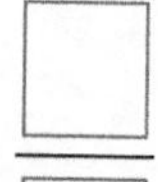
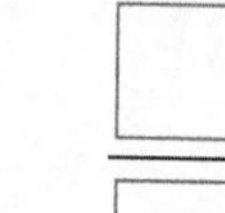
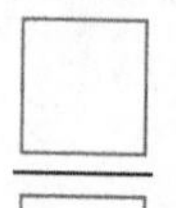
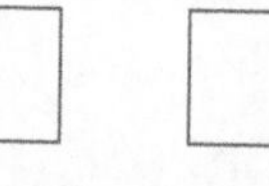

b) $\frac{3}{8}$ $\frac{4}{5}$ $\frac{7}{10}$ $\frac{5}{12}$

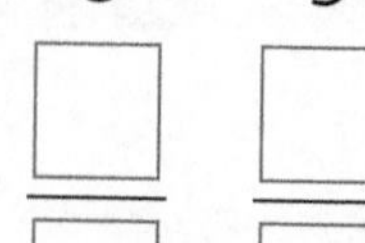
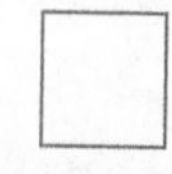
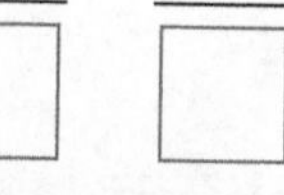
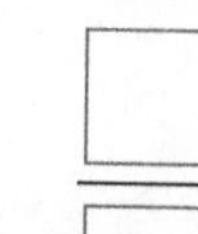
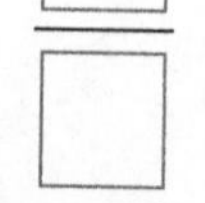
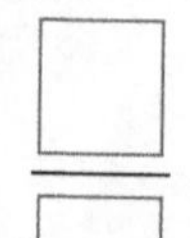

Unit 15: Compare and order fractions, including fractions > 1

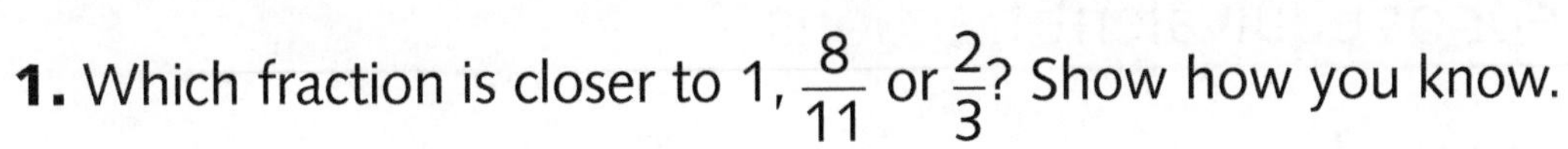

1. Which fraction is closer to 1, $\frac{8}{11}$ or $\frac{2}{3}$? Show how you know.

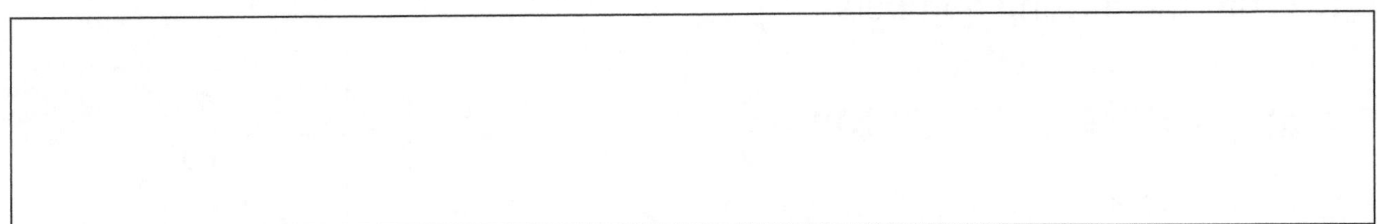

2. Write $> \frac{1}{2}$ or $< \frac{1}{2}$ next to each fraction.

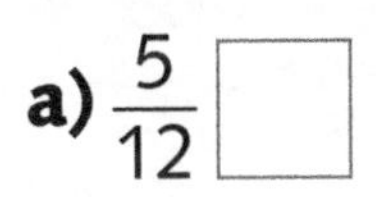

a) $\frac{5}{12}$ ☐

b) $\frac{5}{8}$ ☐

c) $\frac{6}{11}$ ☐

d) $\frac{7}{15}$ ☐

3. Write each fraction in the correct place to make the statement true.

$\frac{7}{9}$ $\frac{4}{15}$ $\frac{2}{3}$ $\frac{2}{5}$

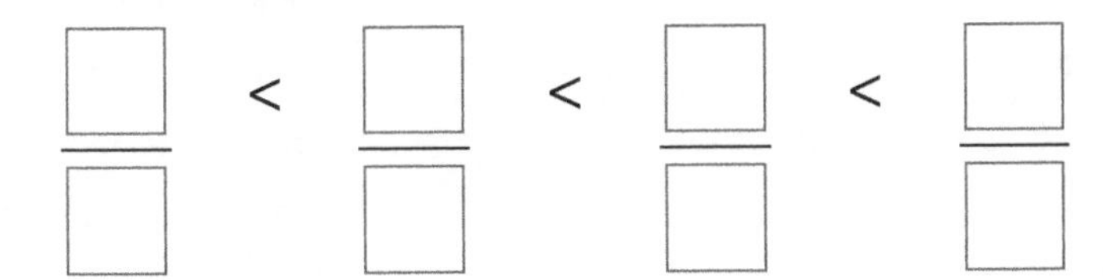

☐/☐ < ☐/☐ < ☐/☐ < ☐/☐

4. Show, using equivalent fractions, that $3\frac{3}{5}$ is less than $\frac{11}{3}$.

Unit 16: Add and subtract fractions with different denominators and mixed numbers, using the concept of equivalent fractions

1. Write the missing numbers.

a) $\frac{3}{7} + \frac{\square}{\square} = 1$ **b)** $1 - \frac{5}{9} = \frac{\square}{\square}$ **c)** $\frac{3}{10} + \frac{\square}{\square} = \frac{7}{10}$

2. Simplify the answer, where possible.

a) $\frac{2}{5} + \frac{1}{10} = \frac{\square}{\square}$ **b)** $\frac{7}{9} - \frac{2}{3} = \frac{\square}{\square}$ **c)** $\frac{1}{8} + \frac{3}{4} = \frac{\square}{\square}$ **d)** $\frac{5}{6} - \frac{1}{12} = \frac{\square}{\square}$

3. Simplify the answer, where possible.

a) $\frac{4}{5} - \frac{2}{3} = \frac{\square}{\square}$ **b)** $\frac{9}{10} - \frac{3}{4} = \frac{\square}{\square}$ **c)** $\frac{7}{9} + \frac{1}{6} = \frac{\square}{\square}$

4. Give your answer as a mixed number.

$\frac{7}{8} + \frac{5}{6} = \frac{\square}{\square}$

5. How much more than $\frac{3}{8}$ is $\frac{11}{12}$? $\frac{\square}{\square}$

6. Fahmida combines $1\frac{3}{4}$ litres of fruit juice with $3\frac{9}{10}$ litres of fizzy water to make a fruit drink. How much fruit drink does Fahmida make?

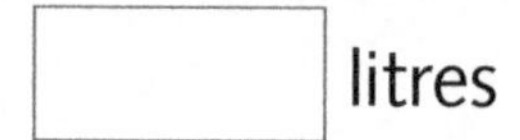

litres

7. Finn ran $4\frac{1}{4}$ kilometres, Ruben ran $3\frac{5}{6}$ kilometres. How much further has Finn run?

☐ kilometres

Unit 16: Add and subtract fractions with different denominators and mixed numbers, using the concept of equivalent fractions

1. Max has completed $\frac{7}{12}$ of his sticker book. What fraction of the book is left to complete?

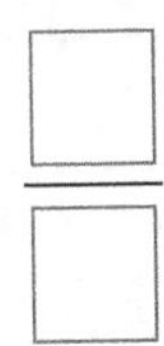

2. Write two fractions to make this calculation correct.

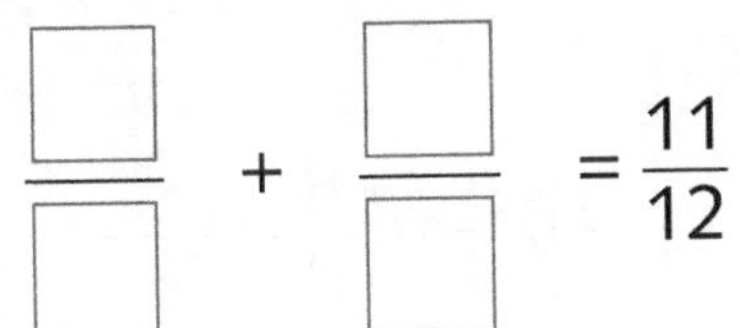

$= \frac{11}{12}$

3. Add the numbers, simplifying each answer, where possible.

a) $\frac{3}{8} + \frac{3}{4} =$

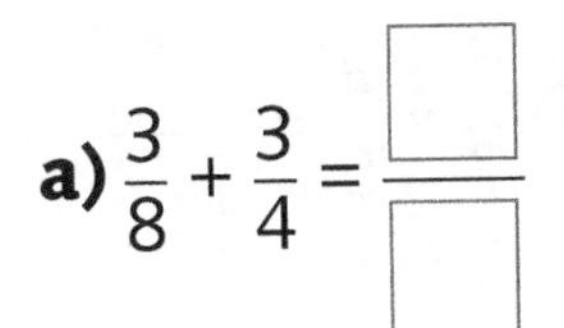

b) $\frac{7}{8} + \frac{2}{3} =$

c) $4\frac{1}{3} + 1\frac{3}{5} =$

4. Work out the difference between the pairs of fractions. Simplify your answers.

a) $\frac{1}{6}$ and $\frac{5}{12}$

b) $\frac{5}{8}$ and $\frac{5}{6}$

c) $5\frac{1}{2}$ and $3\frac{1}{3}$

d) $4\frac{1}{8}$ and $2\frac{1}{6}$

5. Work out: $\frac{1}{2} + \frac{1}{3} + \frac{1}{4} =$

Unit 17: Multiply simple pairs of proper fractions, writing the answer in its simplest form

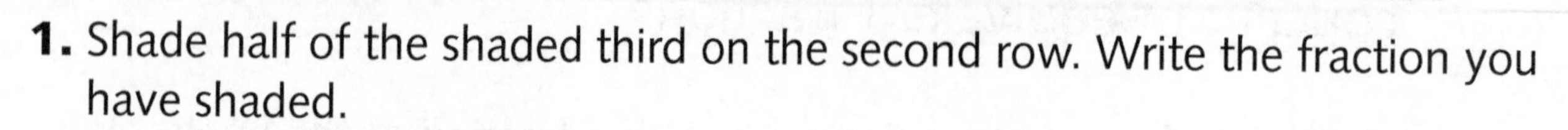

1. Shade half of the shaded third on the second row. Write the fraction you have shaded.

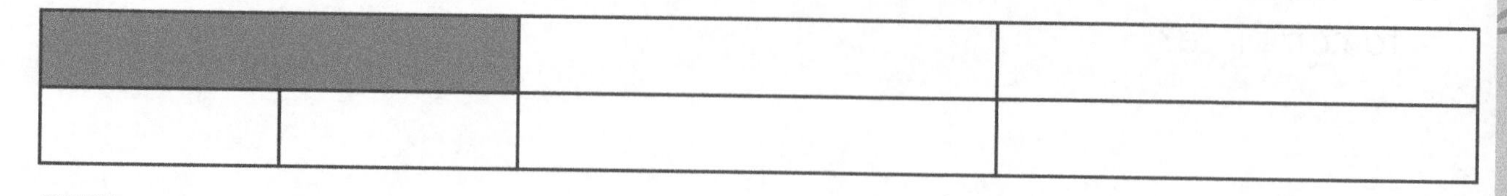

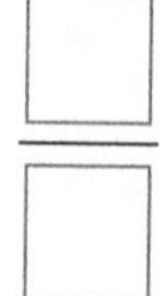

2. a) $\frac{1}{2}$ of $\frac{1}{5}$ = $\frac{\square}{\square}$ **b)** $\frac{1}{2}$ of $\frac{1}{4}$ = $\frac{\square}{\square}$ **c)** $\frac{1}{3}$ of $\frac{3}{10}$ = $\frac{\square}{\square}$ **d)** $\frac{1}{4}$ of $\frac{8}{9}$ = $\frac{\square}{\square}$

3. a) $\frac{1}{6} \times \frac{1}{5} = \frac{\square}{\square}$ **b)** $\frac{1}{10} \times \frac{1}{8} = \frac{\square}{\square}$ **c)** $\frac{3}{4} \times \frac{3}{5} = \frac{\square}{\square}$ **d)** $\frac{2}{7} \times \frac{4}{5} = \frac{\square}{\square}$

4. Give each answer as a simplified fraction.

a) $\frac{3}{8} \times \frac{4}{9} = \frac{\square}{\square}$ **b)** $\frac{5}{8} \times \frac{7}{10} = \frac{\square}{\square}$ **c)** $\frac{2}{3} \times \frac{6}{7} = \frac{\square}{\square}$

5. Write two proper fractions to make the calculation true.

$\frac{\square}{\square} \times \frac{\square}{\square} = \frac{4}{15}$

6. Which is larger, $\frac{3}{4} \times \frac{4}{5}$ or $\frac{4}{5} \times \frac{4}{3}$? Explain your answer.

____

____

$\frac{\square}{\square}$

Unit 17: Multiply simple pairs of proper fractions, writing the answer in its simplest form

1. What is $\frac{1}{2}$ of $\frac{1}{7}$? $\frac{\square}{\square}$

2. Answer these.

a) $\frac{1}{6} \times \frac{1}{3} = \frac{\square}{\square}$

b) $\frac{3}{8} \times \frac{3}{4} = \frac{\square}{\square}$

3 Give the answers as simplified fractions.

a) $\frac{2}{9} \times \frac{3}{5} = \frac{\square}{\square}$

b) $\frac{7}{10} \times \frac{5}{6} = \frac{\square}{\square}$

4. Write the missing digits to make these correct.

a) $\frac{\square}{\square} \times \frac{2}{3} = \frac{10}{21}$

b) $\frac{1}{\square} \times \frac{5}{\square} = \frac{5}{24}$

5. Tick the correct statement.

When a proper fraction is multiplied by another proper fraction, the answer is:

always more than 1

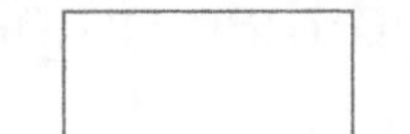

always less than 1

sometimes more than 1, sometimes less than 1 $\square$

Unit 18: Divide proper fractions by whole numbers

1. Use the bars to show $\frac{1}{2} \div 4$ and write the answer.

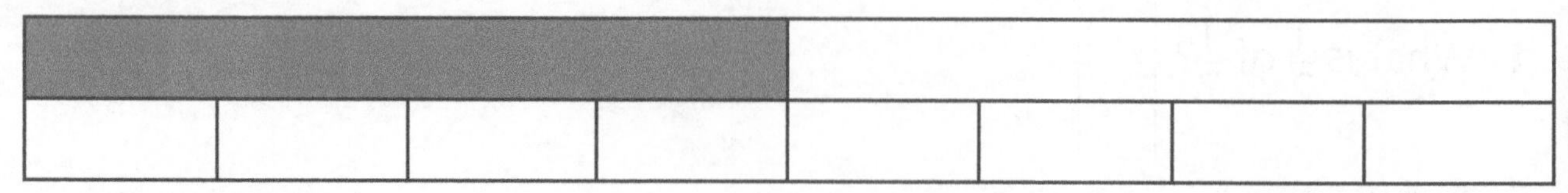

$\frac{1}{2} \div 4 = \frac{\square}{\square}$

2. Use the diagrams to calculate:

a) $\frac{2}{3} \div 2 = \frac{\square}{\square}$

b) $\frac{3}{4} \div 3 = \frac{\square}{\square}$

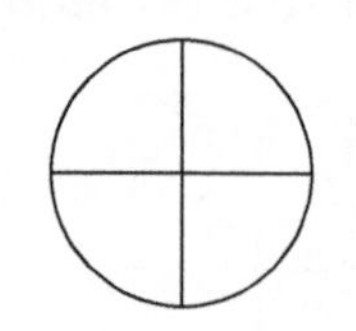

b) $\frac{9}{10} \div 3 = \frac{\square}{\square}$

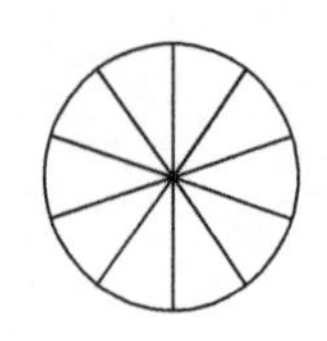

d) $\frac{4}{5} \div 2 = \frac{\square}{\square}$

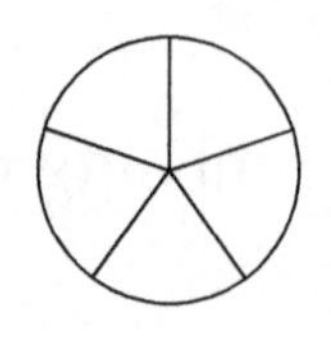

3. Calculate.

a) $\frac{1}{6} \div 3 = \frac{\square}{\square}$ **b)** $\frac{1}{5} \div 4 = \frac{\square}{\square}$ **c)** $\frac{1}{10} \div 5 = \frac{\square}{\square}$

d) $\frac{2}{3} \div 5 = \frac{\square}{\square}$ **e)** $\frac{4}{5} \div 3 = \frac{\square}{\square}$ **f)** $\frac{5}{9} \div 2 = \frac{\square}{\square}$

4. Calculate and then simplify your answers.

a) $\frac{3}{5} \div 9 = \frac{\square}{\square}$ **b)** $\frac{4}{5} \div 8 = \frac{\square}{\square}$

5. Ali and Sarah shared $\frac{3}{4}$ of a pizza equally. What fraction of the pizza did they each have?

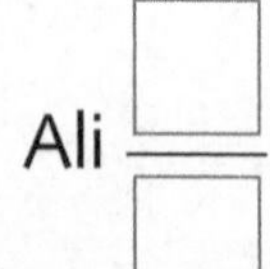

Ali $\frac{\square}{\square}$

Sarah $\frac{\square}{\square}$

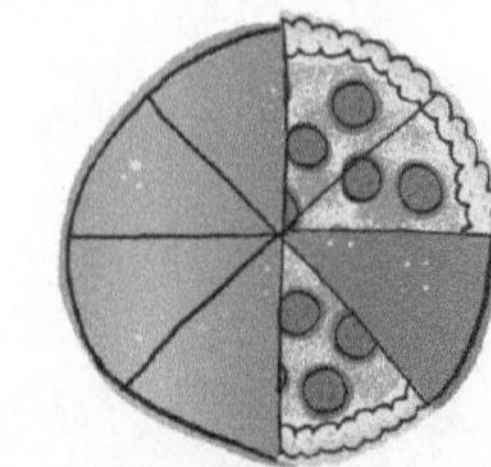

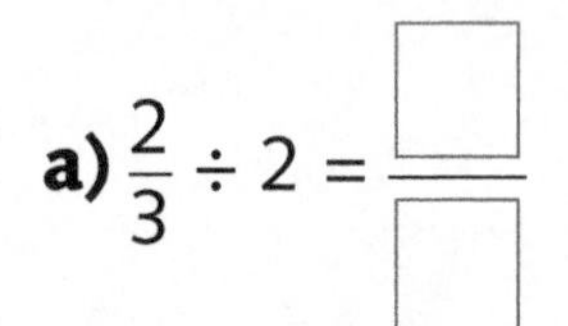

Unit 18: Divide proper fractions by whole numbers

1. Calculate.

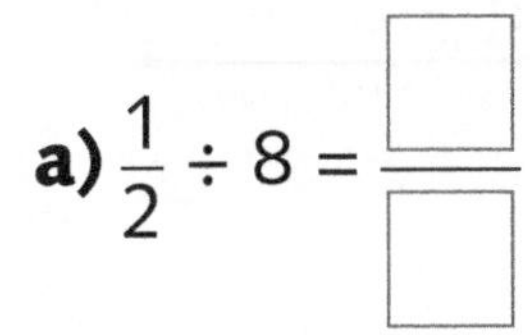

a) $\frac{1}{2} \div 8 = \frac{\square}{\square}$

b) $\frac{1}{6} \div 7 = \frac{\square}{\square}$

2. Calculate.

a) $\frac{5}{6} \div 3 = \frac{\square}{\square}$

b) $\frac{3}{8} \div 5 = \frac{\square}{\square}$

3. Calculate and then simplify your answers.

a) $\frac{4}{9} \div 8 = \frac{\square}{\square}$

b) $\frac{3}{10} \div 6 = \frac{\square}{\square}$

c) $\frac{5}{8} \div 5 = \frac{\square}{\square}$

4. Write numbers in each box to make this division true.

$$\frac{\square}{\square} \div 4 = \frac{1}{\square}$$

5. Three people shared equally $\frac{9}{10}$ of a tin of biscuits. What fraction of the tin did each person get?

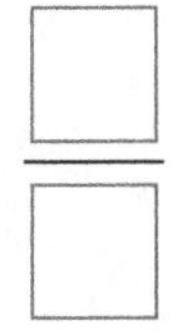

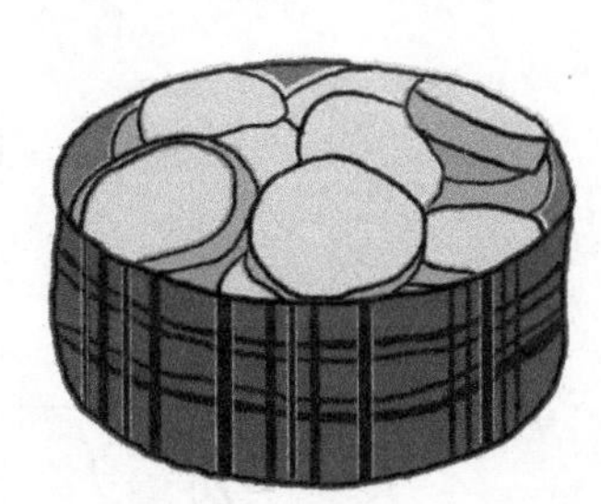

Unit 19: Identify the value of each digit in numbers given to three decimal places and multiply and divide numbers by 10, 100 and 1000 giving answers up to three decimal places

1. What is the value of the digit 3 in these numbers?

a) 42·318 ☐ **b)** 238·409 ☐

c) 21·035 ☐ **d)** 85·973 ☐

2. Write the answers.

a) 26·081 × 100 = ☐ **b)** 48·91 ÷ 100 = ☐

c) 7·05 × 1000 = ☐ **d)** 0·8 ÷ 10 = ☐

3. Write the missing numbers.

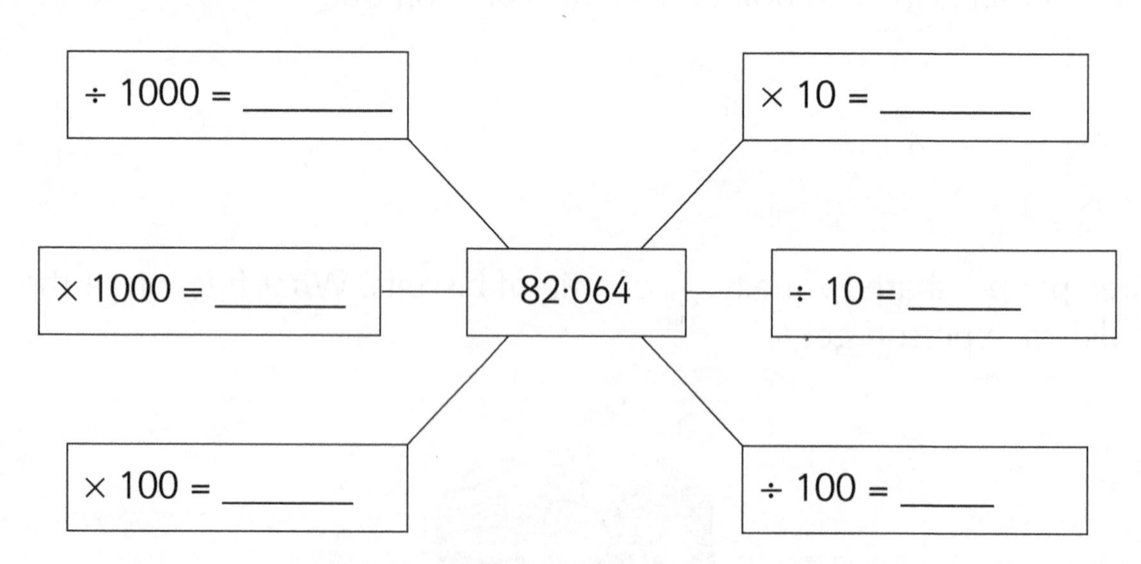

4. a) ☐ × 100 = 176·2 **b)** ☐ ÷ 1000 = 1·935

c) ☐ ÷ 100 = 30·6 **d)** ☐ × 1000 = 467

Unit 19: Identify the value of each digit in numbers given to three decimal places and multiply and divide numbers by 10, 100 and 1000 giving answers up to three decimal places

1. Order these numbers, from smallest to largest.

3·276 3·72 3·3 3·719 3·267

☐ ☐ ☐ ☐ ☐

2. Write × 10, ÷ 10, × 100, ÷ 100, × 1000 or ÷ 1000, to make these correct.

a) 69·045 ☐ = 6904·5

b) 36·1 ☐ = 0·0361

c) 27·43 ☐ = 0·2743

d) 0·093 ☐ = 93

3. Calculate.

a) 27·09 × 1000 = ☐

b) 5·4 ÷ 100 = ☐

c) 0·09 × 100 = ☐

d) 0·04 ÷ 10 = ☐

4. When Grandad John first started work, his daily wage was £12.45. He now earns 10 times that amount. What is his daily wage now?

£ ☐

Unit 20: Multiply one-digit numbers with up to two decimal places by whole numbers

1. Estimate, then calculate, using a column method.

Estimate: ☐ × 4 = ☐ Estimate: ☐ × ☐ = ☐

a) 8·72
× 4

b) 3·06
× 8

2. Estimate, then calculate, using a column method.

a) 6·53 × 9

Estimate:

b) 9·85 × 6

Estimate:

c) 3·94 × 7

Estimate:

3. One packet of cereal costs £3.75. How much will five packets of the same cereal cost?

Estimate: ☐ × ☐ = ☐

4. A plumber wants eight identical sections of pipe, each measuring 2·19 m. What is the total length of the eight pipes?

Estimate: ☐ × ☐ = ☐

Unit 20: Multiply one-digit numbers with up to two decimal places by whole numbers

1. Estimate, then calculate, 7·29 × 4 using a column method.

Estimate: ☐ × ☐ = ☐

2. Estimate, then calculate, 7·26 × 5 using a column method.

Estimate: ☐ × ☐ = ☐

3. A day's bus ticket costs £4.89.
How much is it for six tickets?

Estimate: ☐ × ☐ = ☐

4. Which is more: £8.23 × 9 or £9.23 × 8? Show your working.

Unit 21: Use written division methods in cases where the answer has up to two decimal places

1. Calculate, writing the remainder as a decimal.

a) 3757 ÷ 5 **b)** 7253 ÷ 4

c) 7319 ÷ 8 **d)** 4832 ÷ 3

2. Calculate, writing the remainder as a decimal.

a) 3840 ÷ 25 = ☐ **b)** 4887 ÷ 18 = ☐

3. Six friends share the cost of a take away meal. The total is £25.68. How much do they each pay?

£ 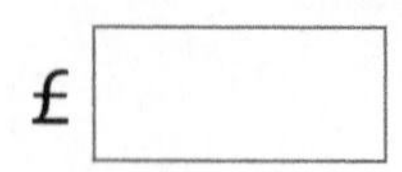

Unit 21: Use written division methods in cases where the answer has up to two decimal places

1. Calculate, writing the remainder as a decimal.

a) 8375 ÷ 4 = ☐

b) 5867 ÷ 9 = ☐

2. Calculate 37·56 ÷ 4.

3. A large holiday cottage costs £2150 to rent. A group of 8 people rent it and share the cost. How much does each person pay?

£ ☐

Unit 22: Associate a fraction with division and calculate decimal fraction equivalents for a simple fraction

1. Write **terminating** or **recurring** next to each of these decimals.

a) 0·55555 ____________ **b)** 0·75 ____________

c) 0·165 ____________ **d)** 0·090909 ____________

2. Draw a line to match each fraction with its decimal equivalent.

$\frac{2}{3}$ $\frac{3}{4}$ $\frac{3}{5}$ $\frac{3}{8}$ $\frac{1}{9}$ $\frac{7}{10}$

0·375 $0{\cdot}\dot{1}\dot{1}\dot{1}$ $0{\cdot}\dot{6}\dot{6}\dot{6}$ 0·7 0·75 0·6

3. Find the decimal equivalent of these fractions, using short division **up to 3 decimal places**.

a) $\frac{1}{8} =$ **b)** $\frac{2}{9} =$

c) $\frac{7}{8} =$

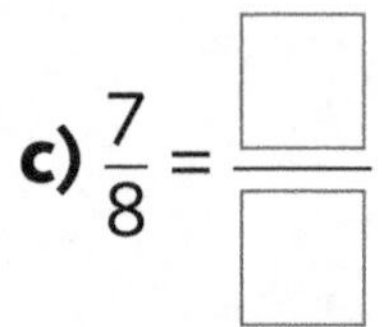

4. Use $\frac{3}{20} = 0{\cdot}15$ to find the decimal equivalent of $\frac{1}{20}$.

5. Calculate these.

a) $\frac{1}{6} \times 2400 =$ **b)** $\frac{1}{10}$ of 2860 =

c) $\frac{8}{9}$ of 720 = **d)** $\frac{5}{8} \times 4800 =$

e) $\frac{7}{10} \times 380 =$ **f)** $\frac{3}{4}$ of 832 =

Unit 22: Associate a fraction with division and calculate decimal fraction equivalents for a simple fraction

1. Write an example of each of the following:

a) a recurring decimal with two repeated digits

b) two fractions with a terminating decimal equivalent

c) two fractions with a recurring decimal equivalent

2. Use short division to convert these fractions to decimals.

a) $\frac{1}{6}$ **b)** $\frac{5}{8}$

3. Use the decimal equivalents to show that $\frac{3}{10} < \frac{1}{3}$.

4. $\frac{5}{9}$ of 7200 =

5. $\frac{3}{10} \times 6200$ =

Unit 23: Solve problems which require answers to be rounded to specified degrees of accuracy

1. There are 215 children at a winter sports camp.

a) How many complete teams of 7 for netball can be formed?

[ ] teams

b) How many teams of 11 for football can be formed?

[ ] teams

c) How many teams of 15 for rugby can be formed?

[ ] teams

2. The whole of Key Stage 2 is going on a trip using minibuses that carry 15 people each. How many minibuses will be needed to transport 125 children and 5 adults?

[ ] minibuses

3. Shilpa collects stickers and puts them in an album, 9 to a page. She has 125 stickers. How many pages has she filled?

[ ] pages

4. The number of people living in a large city, rounded to the nearest 100 000, is 300 000. What is the **most** number of people that could live in the city?

[ ] people

5. Scissor blocks hold 12 pairs of scissors. In the art room, there are 50 pairs of scissors. How many scissor blocks does the school need to buy?

[ ] scissor blocks

Unit 23: Solve problems which require answers to be rounded to specified degrees of accuracy

1. How many 3 m lengths of fabric can be cut from a 20 m roll?

2. What is the nearest multiple of 9 to the number 85?

3. Six friends share the cost of a birthday present for a friend. The present cost £35. How much does each person need to pay, to the nearest pence, to cover the cost?

£

4. Mollie has saved £5.80 to spend at a second hand book fair where every book is 25p. How many books can she buy?

books

5. School pens are sold in packs of 24. How many packs are needed for each of the 476 schoolchildren to have a pen?

packs

Unit 24: Recall and use equivalences between simple fractions, decimals and percentages, including in different contexts

1. Complete the table to show the equivalent fractions, decimals and percentages. Use the working out box to find the answers, using the clues given. Simplify the fractions.

Fraction	Decimal	%
N ÷ D →	→ × 100 →	
		90%
$\frac{2}{3}$		
	$0{\cdot}\dot{7}$	
		$33{\cdot}\dot{3}$
$\frac{1}{8}$		
	0·65	
		3%

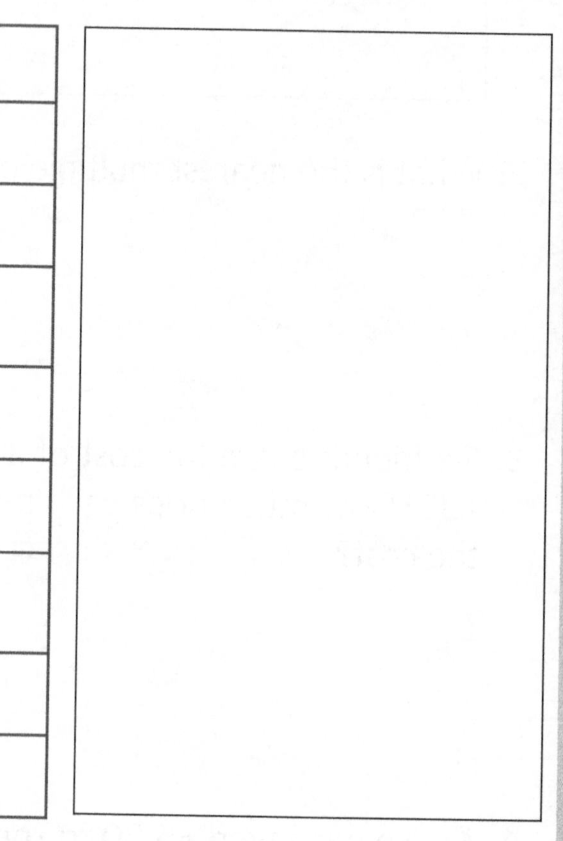

2. Finn and Ruben receive the same amount of pocket money. Finn saved 35% of his and Ruben saved $\frac{3}{8}$ of his. Who saved the most? Show how you know.

3. Write these fractions, decimals and percentages on the number line.

a) $\frac{3}{10}$ **b)** 0·48 **c)** 36% **d)** $\frac{3}{5}$

Unit 24: Recall and use equivalences between simple fractions, decimals and percentages, including in different contexts

1. Look at the number line. Suggest a fraction at A, a decimal at B and a percentage at C.

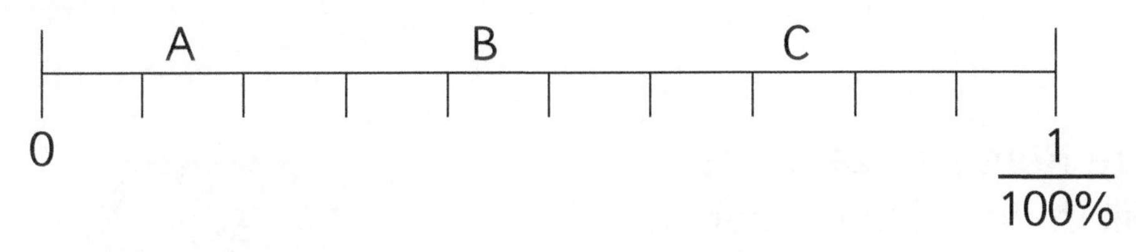

A = 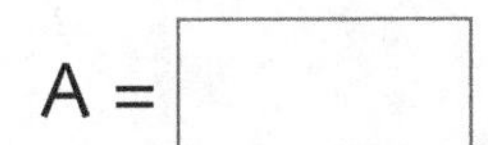 B = ☐ C = ☐

2. Write each amount as a fraction, decimal or percentage. Simplify the fractions.

a) $\frac{2}{9}$ = 0· ☐ = ☐ %

b) 0·15 = ☐ % = $\frac{☐}{☐}$ = $\frac{☐}{☐}$

c) 84% = 0· ☐ = $\frac{☐}{☐}$ = $\frac{☐}{☐}$

3. Order these fractions, decimals and percentages, from smallest to largest.

0·45 $\frac{2}{5}$ $\frac{1}{3}$ 0·32 42%

4. In an election for the school council, Lisa received 25% of the vote, David received $\frac{2}{5}$ and Zac received the rest of the votes. Who received the most votes? Show how you know.

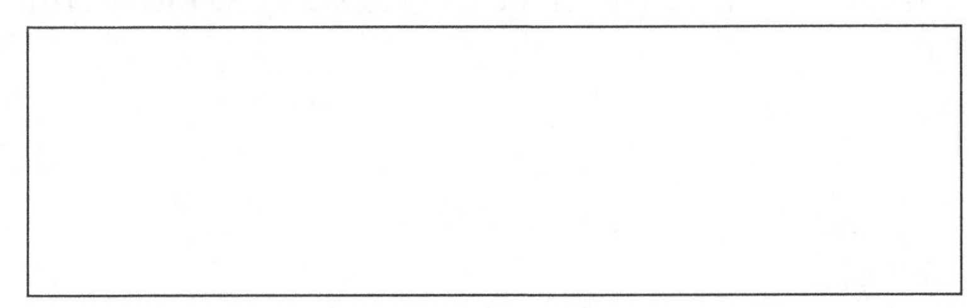

Unit 25: Solve problems involving the relative sizes of two quantities where missing values can be found by using integer multiplication and division facts

1. Eight pens cost £6.40. How much are 5 pens?

£ ☐

2. These ingredients are from a pizza recipe serving 4 people. Fill in the table to change the servings and answer the questions.

240 g bread dough
80 ml chopped tomatoes
160 g grated cheese

Serves	Dough	Tomatoes	Cheese
4	240 g	80 ml	160 g

a) How much bread dough is needed for 8 people? ☐ g

b) How much tomato is needed for 6 people? ☐ ml

c) How much cheese is needed for 1 person? ☐ g

d) Tom only has 60 ml of tomatoes. How much cheese and dough should he use?

cheese ☐ g dough ☐ g

3. 250 g of rolled oats costs 76p. How much is that for 1 kg? ☐

4. A coffee shop uses 125 ml of milk for each milky coffee it serves. How many coffees will 2 litres of milk make?

☐ coffees

Unit 25: Solve problems involving the relative sizes of two quantities where missing values can be found by using integer multiplication and division facts

1. Here are the ingredients to make 12 blueberry muffins.

1 egg
120 ml milk
60 ml vegetable oil
240 g plain flour
120 g caster sugar
2 teaspoons baking powder
150 g blueberries

a) How much flour and blueberries are needed to make 24 muffins?

flour [ ] g blueberries [ ] g

b) How much vegetable oil and baking powder are needed for 6 muffins?

vegetable oil [ ] ml baking powder [ ] teaspoons

c) How much milk would be used to make 18 muffins? [ ] ml

d) How much sugar is there in 2 muffins? [ ] g

2. The average mass of 1 orange is twice that of an apple. An average apple weighs 150 g. What is the total mass of 2 apples and 3 oranges?

[ ] g

3. A large packet of cereal has a mass of 500 g. Tanya has a 25 g portion each day. How many days will the packet last her?

[ ] days

Unit 26: Solve problems involving the calculation of percentages

1. Calculate.

a) 10% of 180 = ☐ **b)** 50% of 284 = ☐

c) 20% of 330 = ☐ **d)** 15% of 1400 = ☐

e) 95% of 280 = ☐ **f)** 1% of 640 = ☐

2. The pie chart shows the approximate percentage of the three main food groups for a balanced diet. Estimate the percentage of each section.
Make sure the total is 100%.

carbohydrates  %

fats  %

proteins 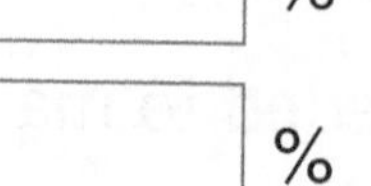 %

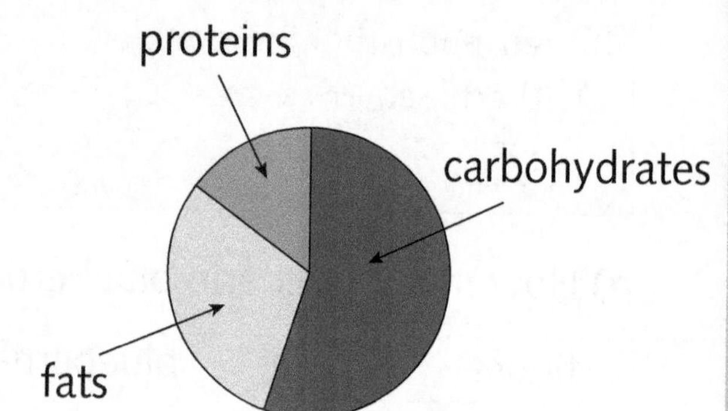

3. In a class, 30% are boys and the rest are girls. There are 21 girls. How many children are in the class?

☐ children

4. Answer the questions, using the bar models to help.

a) 10% of a number is 7. What is the number?

b) 40% of a number is 12. What is the number?

c) 75% of a number is 180. What is the number?

Unit 26: Solve problems involving the calculation of percentages

1. Calculate.

a) 20% of 450 = ☐

b) 15% of 320 = ☐

c) 95% of 1600 = ☐

2. In a school of 250 pupils, 40% are boys. How many of the pupils are girls?

☐

3. Answer the questions.

a) 30% of a number is 33. What is the number? ☐

b) 25% of a number is 150. What is the number? ☐

4. The pie chart shows how children travel to school. 200 children took part in the survey.

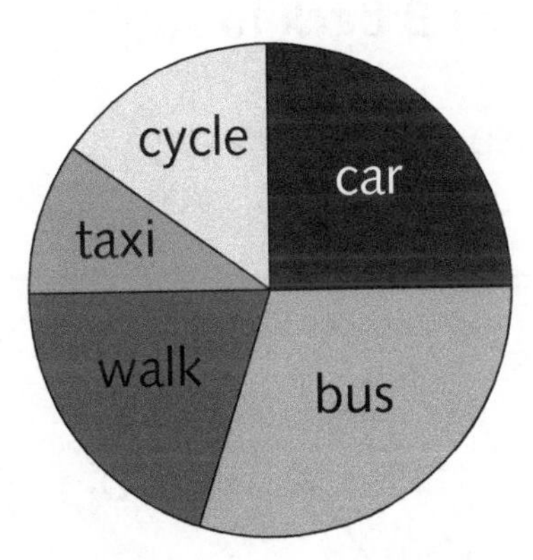

a) How many children travel by car? ☐ children

b) Estimate **the percentage** of children who come to school by bus. ☐ %

c) 20% walk to school. Estimate the percentage who cycle. ☐ %

Unit 27: Solve problems involving similar shapes where the scale factor is known or can be found

1.

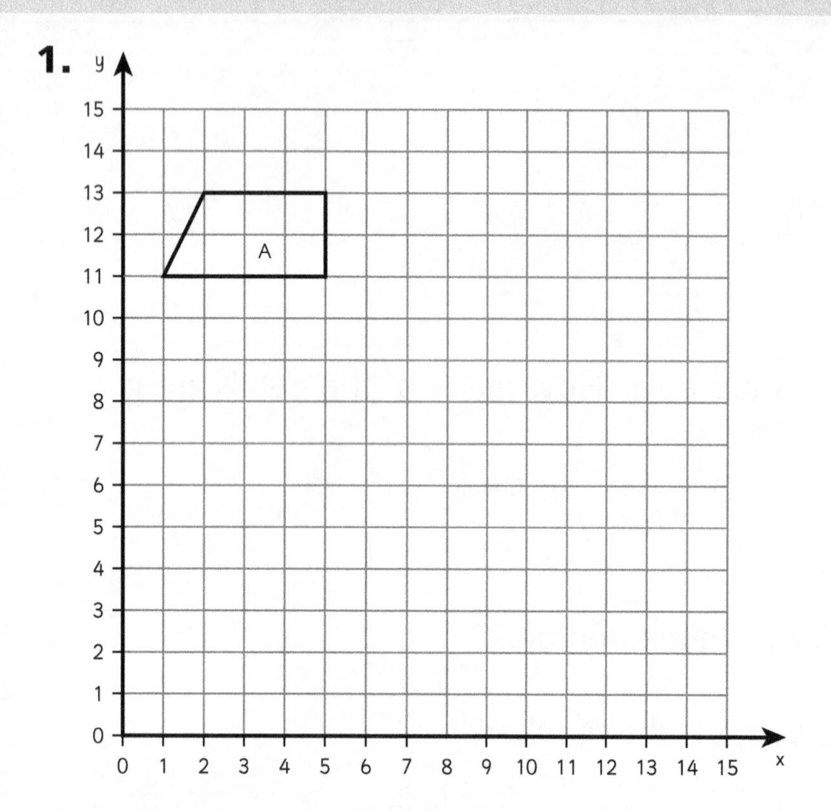

a) Enlarge shape A by scale factor 2. Label the new shape B.

b) What is the scale factor of **B back to A**?

2. Look at the lengths of these similar shapes.

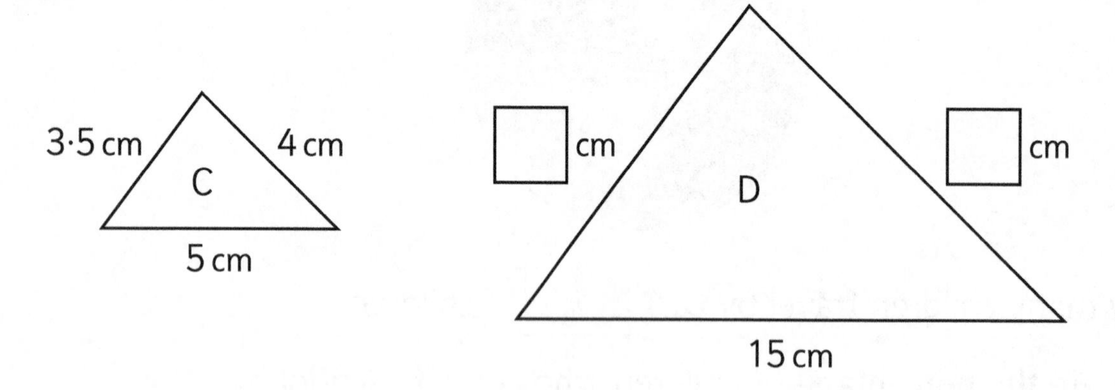

a) What is the scale factor of C to D?

b) Write in the missing lengths of shape D.

Unit 27: Solve problems involving similar shapes where the scale factor is known or can be found

1. Enlarge shape A by scale factor 1·5. Label the new shape B.

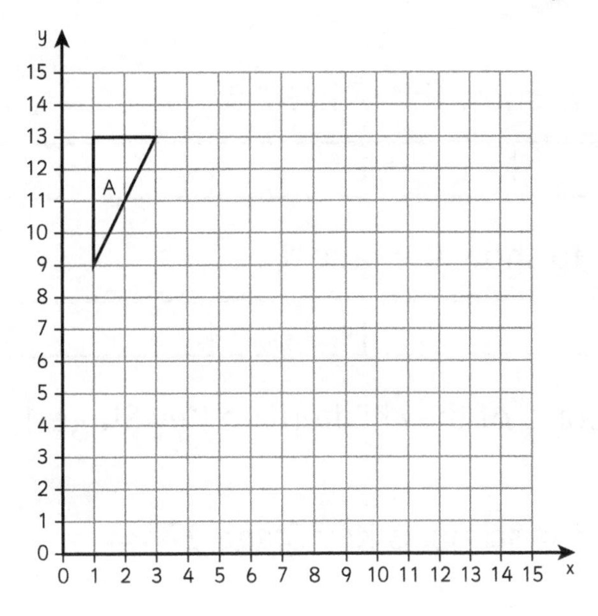

2. a) **b)**

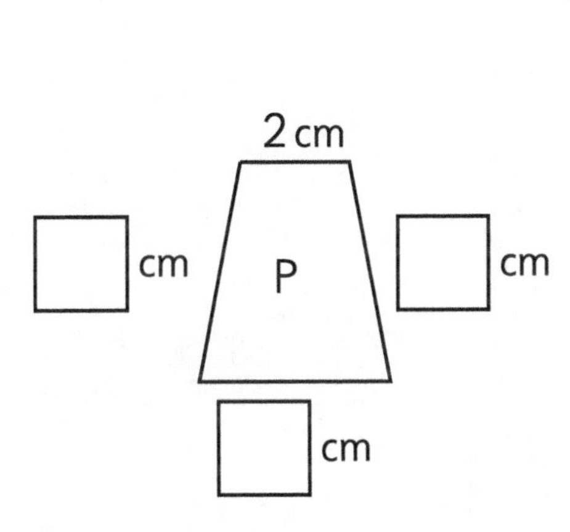

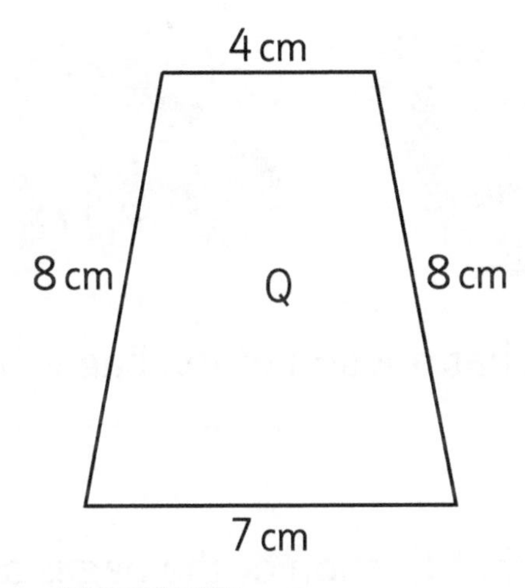

a) By what scale factor has P been enlarged?

b) Write in the missing lengths on shape P.

3. A triangle with sides of 16 cm, 20 cm and 12 cm is enlarged by scale factor $\frac{1}{4}$. What are the lengths of the three sides in the enlarged triangle?

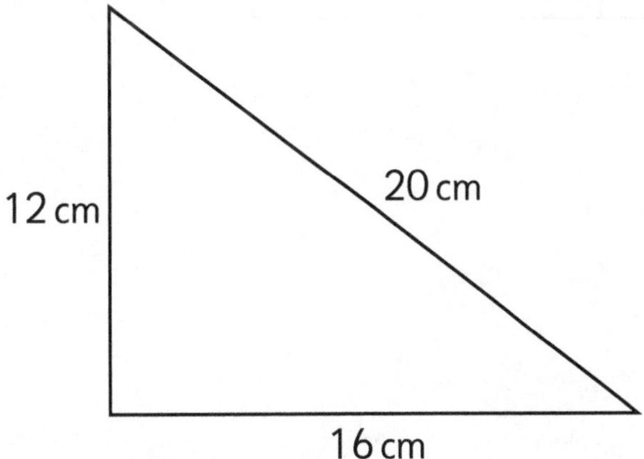

Unit 28: Solve problems involving unequal sharing and grouping using knowledge of fractions and multiples

1. Complete these.

a) Shade the bar to show the ratio 3 : 5.

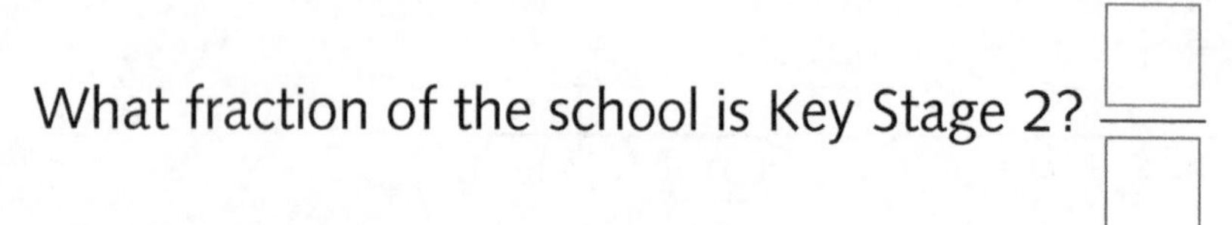

b) Shade the bar to show the ratio 1 : 3.

2. In a Primary school, $\frac{3}{7}$ of the children are Key Stage 1, the rest are Key Stage 2.

What fraction of the school is Key Stage 2? $\frac{\square}{\square}$

3. In a bag of fruits, the ratio of bananas to apples is 4 : 5.

a) What fraction of the bag is bananas? $\frac{\square}{\square}$

b) What fraction of the bag is apples? $\frac{\square}{\square}$

4. In a pattern of coloured tiles, $\frac{1}{4}$ are blue and the rest are white. What is the ratio of blue to white tiles?

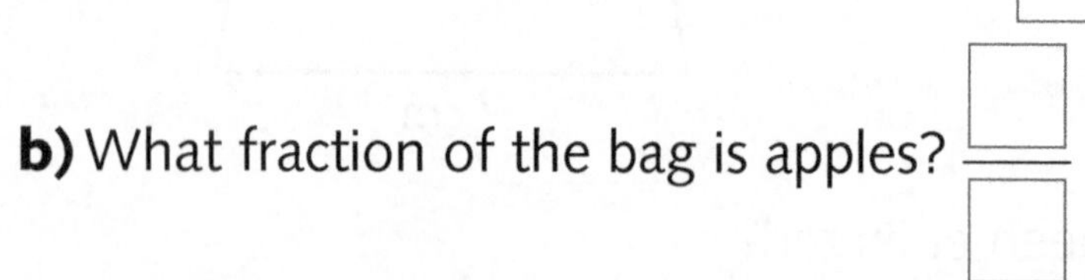

ratio: ☐

Unit 28: Solve problems involving unequal sharing and grouping using knowledge of fractions and multiples

5. Adil read $\frac{3}{8}$ of his book on Friday, and the other 45 pages on Saturday. How many pages are in Adil's book? Use the bar model to help you.

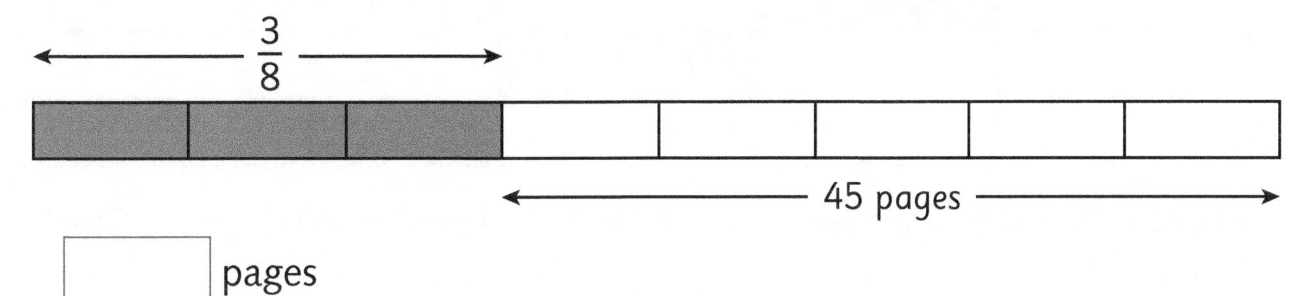

☐ pages

6. Write the pairs of equivalent ratios.

6 : 8   10 : 15   10 : 20   15 : 5

20 : 30   6 : 2   9 : 12   5 : 10

---

7. In a pancake recipe, there is 200 g of flour for every 175 ml of milk. Write the ratio of flour to milk **in its simplest** form.

---

8. Matt and Chloe share 60 stickers in the ratio 3 : 2. How many stickers did they both get?

Matt: ☐ stickers   Chloe: ☐ stickers

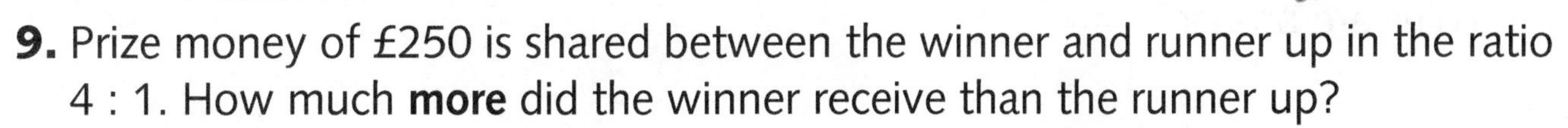

9. Prize money of £250 is shared between the winner and runner up in the ratio 4 : 1. How much **more** did the winner receive than the runner up?

£ ☐

10. Two numbers add together to make 9·6. One number is double the other number. What are the two numbers?

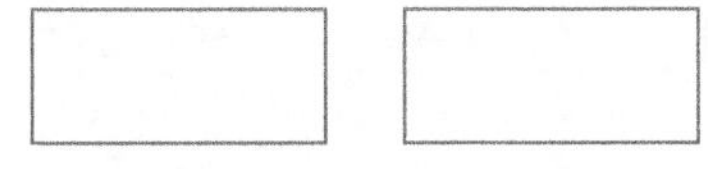

11. Share £120 in the ratio 1 : 2 : 3. How much is each part worth?

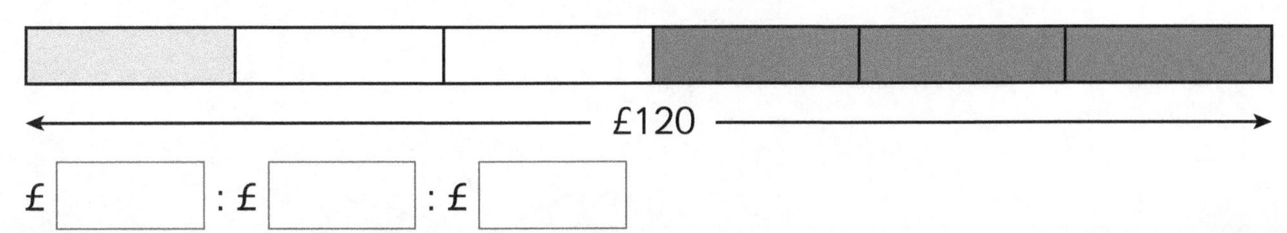

£ ☐ : £ ☐ : £ ☐

Unit 28: Solve problems involving unequal sharing and grouping using knowledge of fractions and multiples

1.

a) Write the ratio of shaded to unshaded cells shown in the diagram. Simplify your answer.

____

b) What is the simplified fraction of the shaded part?

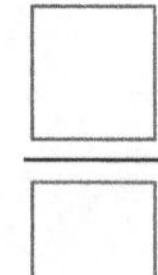

c) What is the simplified fraction of the unshaded part?

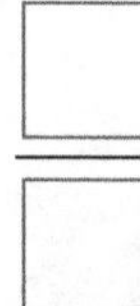

2. In a bag of coins, $\frac{3}{5}$ are silver and the rest are bronze. What fraction of the

coins are bronze?

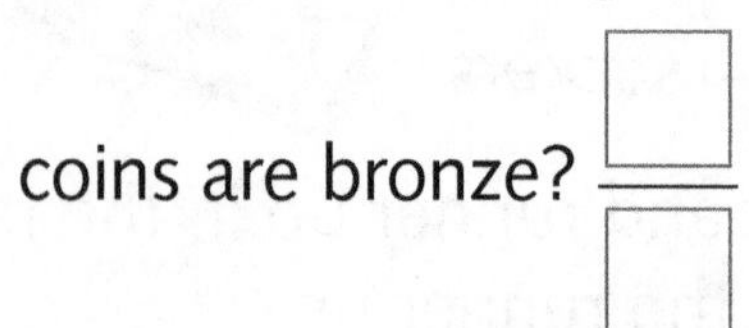

3. In a class, $\frac{1}{6}$ of the children are left-handed. What is the **ratio** of

left- to right-handed children? ____________

4. Simplify these ratios.

a) 2 : 6 ____________________

b) 25 : 15 ____________________

c) 225 : 75 ____________________

Unit 28: Solve problems involving unequal sharing and grouping using knowledge of fractions and multiples

5. Anna has used $\frac{3}{10}$ of a bag of sugar. She has 350 g left. What was the mass of the bag before she used the sugar?

[ ] g

6. Ten satsumas cost £2 and 5 bananas cost £1.50. How much more does a banana cost than a satsuma?

[ ] p

7. A decorator makes purple paint from red and blue paint, using the ratio 2 : 5. He needs to make 21 litres to paint a hall. How much of each colour will he need?

red: [ ] litres blue: [ ] litres

8. The organiser of a school fair needs to buy cartons of orange and apple juice with the ratio of 5 : 4. She needs to buy 360 cartons. How many of orange and how many of apple should she buy?

orange: [ ] cartons apple: [ ] cartons

Unit 29: Express missing number problems algebraically

1. Calculate the value of each symbol.

a) $\phi + 25 = 32$ $\phi =$ ☐ **b)** $47 - \lambda = 21$ $\lambda =$ ☐

c) $5 \times \chi = 35$ $\chi =$ ☐ **d)** $42 \div \Delta = 6$ $\Delta =$ ☐

2. Calculate the value of each letter.

a) $21 - a = 13$ $a =$ ☐ **b)** $6b = 24$ $b =$ ☐

c) $c - 12 = 30$ $c =$ ☐ **d)** $d \div 3 = 4$ $d =$ ☐

3. Find each number.

a) When I multiply my number by 4, then subtract 8, the answer is 32.
What is my number? ☐

b) When I divide my number by 6, then add 7, the answer is 12.
What is my number? ☐

4. The symbol or letter represents the same number on both sides of the equals sign. What value for each symbol makes the equation true?

a) $9 - \Delta = \Delta - 1$ $\Delta =$ ☐

b) $7\phi = 18 + \phi$ $\phi =$ ☐

c) $20 - y = 2y + 2$ $y =$ ☐

5. Find the value of each shape in the puzzle.

○ + ○ = 70

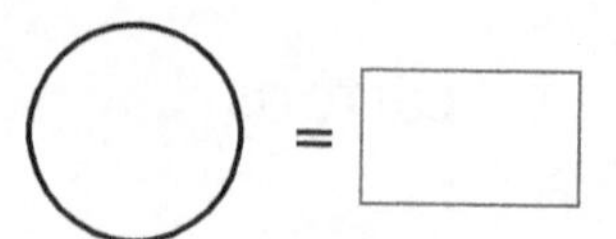

○ + △ = 42

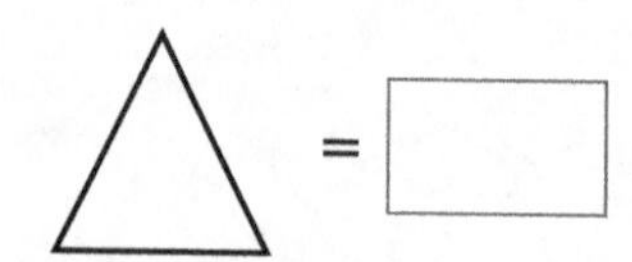

○ + △ + □ = 50

□ = ☐

Unit 29: Express missing number problems algebraically

1. Find the value of n in each equation.

a) $30 - n = 20$ $n =$ ☐

b) $8n = 72$ $n =$ ☐

c) $54 \div n = 9$ $n =$ ☐

d) $36 + n + n = 100$ $n =$ ☐

2. $4t - 18 = 62$. Find the value of t.

$t =$ ☐

3. Solve the number puzzle to find the value of each shape.

smiley face + sun = 10 smiley face = ☐

lightning bolt + smiley face = 21 lightning bolt = ☐

sun + lightning bolt = 15 sun = ☐

Unit 30: Find pairs of numbers that satisfy an equation with two unknowns

1. Write eight solutions for p and q that make $p - q = 10$ true.

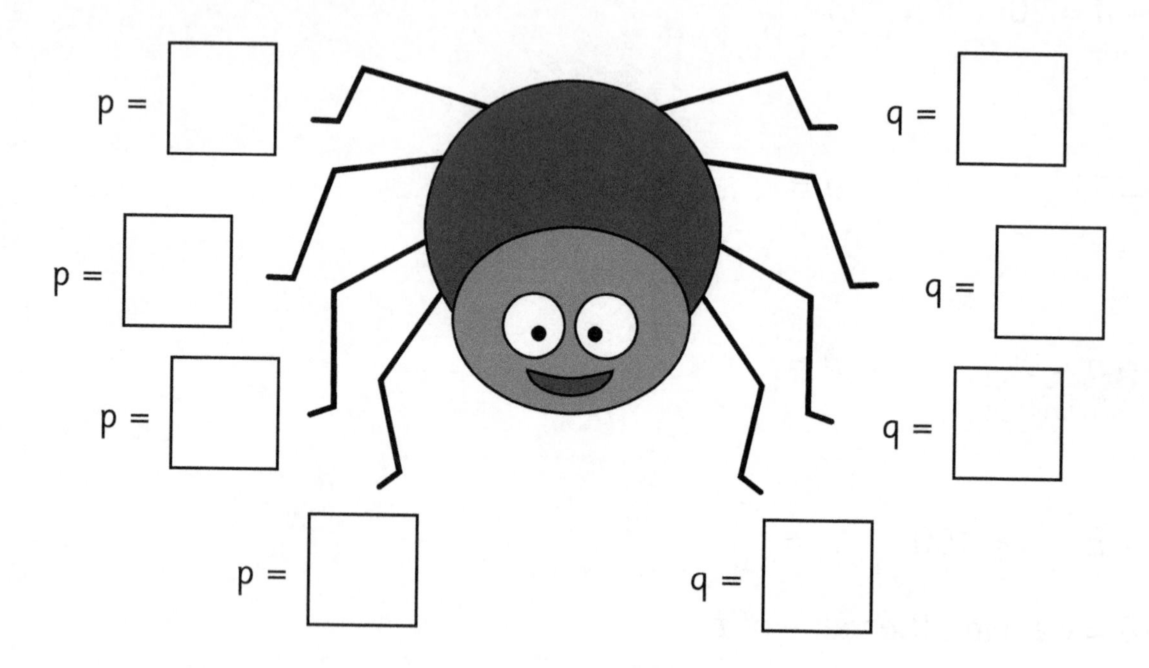

2. If $xy = 30$, what is the value of y when $x = 3$?

y =

3. Do these pairs of values for c and d satisfy the equation $cd = 40$? Write **yes** or **no** against each pair.

a) $c = 4$, $d = 10$ ______________

b) $d = 20$, $c = 20$ ______________

c) $c = 0.8$, $d = 5$ ______________

d) $c = -5$, $d = -8$ ______________

4. Which values for x and y satisfy **both** $x + y = 12$ **and** $x - y = 2$?

x =

y =

Unit 30: Find pairs of numbers that satisfy an equation with two unknowns

1. Write all the positive number solutions for g and h that satisfy the equation $gh = 36$.

$g =$ ☐ $h =$ ☐

$g =$ ☐ $h =$ ☐

$g =$ ☐ $h =$ ☐

$g =$ ☐ $h =$ ☐

$g =$ ☐ $h =$ ☐

$g =$ ☐ $h =$ ☐

2. Write three solutions for the equation $a + b - c = 5$.

$a =$ ☐ $b =$ ☐ $c =$ ☐

$a =$ ☐ $b =$ ☐ $c =$ ☐

$a =$ ☐ $b =$ ☐ $c =$ ☐

3. $a + b = -10$. When $a = 2$, what is the value of b?

$b =$ ☐

4. Which values for e and f satisfy **both** $e + f = 16$ and $ef = 55$?

$e =$ ☐

$f =$ ☐

Unit 31: Enumerate possibilities of combinations of two variables

1. Complete the pairs of **whole-number** values for x **and** y that satisfy the equation $x + 2y = 15$. You can use 0 as one value.

15		
x	y	y

a) $x = 1$, $y =$ ☐ **b)** $x = 3$, $y =$ ☐

c) $x = 5$, $y =$ ☐ **d)** $y = 4$, $x =$ ☐

e) $y = 3$, $x =$ ☐ **f)** $y = 0$, $x =$ ☐

g) Show that $x = 2$ does not give a whole-number answer for y. ______

2. List the six solutions for $2x + y = 11$.

$x =$ ☐ $y =$ ☐ $x =$ ☐ $y =$ ☐

$x =$ ☐ $y =$ ☐ $x =$ ☐ $y =$ ☐

$x =$ ☐ $y =$ ☐ $x =$ ☐ $y =$ ☐

3. Here are some ice-cream toppings. Make a list of all the combinations of **two different** toppings that can be made. Use the letters in brackets to record the combinations. For example, SC for sauce/chocolate.

Sauce (S) Chocolate (C) Fruit (F)

Marshmallows (M) Nuts (N)

______ ______

______ ______

______ ______

______ ______

______ ______

Unit 31: Enumerate possibilities of combinations of two variables

1. Show that $x = 5$ and $y = 11$ is a solution to $3x - y = 4$.

2. List six solutions for $5x - y = 0$.

$x =$ ☐ $y =$ ☐ $x =$ ☐ $y =$ ☐

$x =$ ☐ $y =$ ☐ $x =$ ☐ $y =$ ☐

$x =$ ☐ $y =$ ☐ $x =$ ☐ $y =$ ☐

3. Look at your answers to Question 2. What is special about the values for y?

____

4. Here is a menu. List all the combinations of a drink with a snack. Use the letters in brackets for your list. One combination has been written for you.

Drinks	**Snacks**	
Hot chocolate (H)	Fruit (F)	HF
Milk (M)	Crisps (C)	________
Juice (J)	Biscuit (B)	________
		________
		________
		________
		________
		________
		________
		________

Unit 32: Generate and describe linear number sequences

1. Write the next five terms in these number sequences and the term-to-term rule.

a) 5, 10, 15, 20, ☐ ☐ ☐ ☐ ☐ rule ____

b) 80, 72, 64, 56, ☐ ☐ ☐ ☐ ☐ rule ____

c) 19, 16, 13, 10, ☐ ☐ ☐ ☐ ☐ rule ____

2. Use the linear sequence to answer the questions.

5, 8, 11, 14, 17…

a) What is the 1st term in this sequence? ☐

b) What is the term-to-term rule for the sequence? ____________

c) Write the next three terms in the sequence. ☐ ☐ ☐

3. Generate the first five terms of these sequences.

a) Start at 1, add 9.

☐ ☐ ☐ ☐ ☐

b) Start at 60, subtract 7.

☐ ☐ ☐ ☐ ☐

c) Start at 2, add 15.

☐ ☐ ☐ ☐ ☐

d) Start at 10, subtract 4.

☐ ☐ ☐ ☐ ☐

4. Will **50** be a term in these sequences? Write yes or no, explaining how you know.

a) 105, 95, 85, 75… ____________

b) 8, 14, 20, 26… ____________

Unit 32: Generate and describe linear number sequences

1. Describe the following sequence:

3, 8, 13, 18…

a) 1st term: ☐

b) Increasing or decreasing? ____________________

c) Term-to-term rule: ____________________

d) Next two terms: ☐ ☐

2. Generate the first five terms of these sequences.

a) Start at 100, subtract 30.

☐ ☐ ☐ ☐ ☐

b) Add 1 to the first five multiples of 8.

☐ ☐ ☐ ☐ ☐

3. Fill in the missing numbers in these sequences.

a) ☐ 58 49 ☐ 31 ☐

b) 91 ☐ ☐ ☐ ☐ 46

4.

a) Draw the next pattern in the picture sequence.

b) Write the number of circles as a number sequence.

☐ ☐ ☐ ☐

c) How many circles will be in the 5th picture? ☐

Unit 33: Use simple formulae

1. Draw a line to connect the formula to the correct area or perimeter.

$4s$	area of a rectangle
s^2	volume of a cube
$2(w + l)$	perimeter of a square
wl	area of a square
s^3	perimeter of a rectangle

2. The area of a triangle is half the base (b) times height (h).

a) Complete the formula for the area of a triangle.

A = ____ × ____ ÷ ____

b) Use the formula to work out the area of a triangle with a base of 10 cm and height of 7 cm.

☐ cm²

3. Use the formula D = ST to:

a) calculate the value of D when S = 6 and T = 1·5

D = ☐

b) calculate the value of S when D = 150 and T = 3

S = ☐

4. From an internet site, posters cost £4 each, plus a standard delivery charge of £3, no matter how many posters you buy.

a) How much will it cost for 5 posters, including delivery? £ ☐

b) Sarah spent £39. How many posters did she buy? ☐ posters

Unit 33: Use simple formulae

1. The formula to work out the perimeter, P, of a regular hexagon with sides, s, is $P = 6s$.

a) Work out the perimeter of a regular hexagon with sides of length 5 cm.

[ ] cm

b) The perimeter of a regular hexagon is 60 cm. What is the length of each side?

[ ] cm

2. Some families are going to the cinema. Adult (A) tickets cost £7 and children's (C) tickets cost £4.

a) Circle the correct formula for the total cost, T, of a mix of adult and children's tickets.

$T = 7C + 4A$ $7T = A + 4C$ $T = 7A + 4C$ $T = 7A - 4C$

b) Work out the total cost for 4 adults and 2 children. £ [ ]

c) A family paid £34 to take 5 children and some adults to the cinema. How many adults went with the children?

[ ] adults

3. Use the formula $R = 2x + y$:

a) to find the value of R when $x = 3$ and $y = 2$

$R =$ [ ]

b) to find the value of x when $R = 22$ and $y = 10$.

$x =$ [ ]

Unit 34: Read, write and convert between standard units of length, mass, volume and time, using up to three decimal places

1. Complete the tables to convert between:

a) litres and ml

b) cm and metres.

litres	ml
2	
	3500
6·7	
	750
2·25	
	600
0·1	
	90

cm	metre
300	
	4·5
2860	
	0·9
45	
	0·04
7	
	1·08

2. Answer these questions related to time.

a) What is 4·25 hours in minutes?

☐ minutes

b) Show whether 3000 seconds is more or less than 1 hour.

____

c) What is 425 minutes in hours and minutes?

☐ hours ☐ minutes

3. Write < or > between these pairs of measures to compare them.

a) 3·4 kg ☐ 3399 g

b) 105 mm ☐ 11 cm

c) 7·02 m ☐ 710 cm

d) 245 secs ☐ 6 minutes

Unit 34: Read, write and convert between standard units of length, mass, volume and time, using up to three decimal places

1. Complete these conversions between metric units.

a) 3650 m = ☐ km

b) 3·05 m = ☐ cm

c) 0·4 cm = ☐ mm

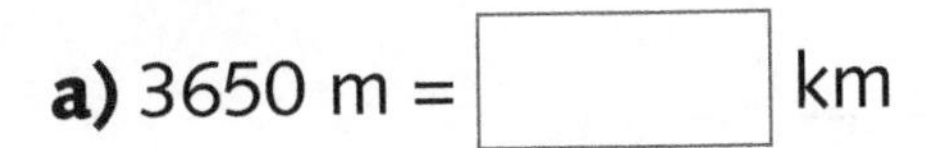

2. Mark, with an arrow, 650 ml on the measuring jug.

3. Which is longer, 465 minutes or 7·5 hours? Show your working.

4. Order these lengths, from longest to shortest.

13·5 metres   0·3 km   1354 cm   13 505 mm

Unit 35: Convert between miles and kilometres

1. Circle the two correct conversions for miles and kilometres.

1·6 miles = 1 km   5 miles = 8 km   8 km = 5 miles   1 mile = 1·6 km

2. Complete the tables.

miles	km
10	16
5	
	32
50	
	160
250	
	480

miles	km
10	16
1	
	3·2
0·5	
	6·4
2·5	
	64

3. The signpost shows that the distance to London is 150 miles. A visitor from Spain is confused, as she is used to kilometres. Find the equivalent distance to London in kilometres.

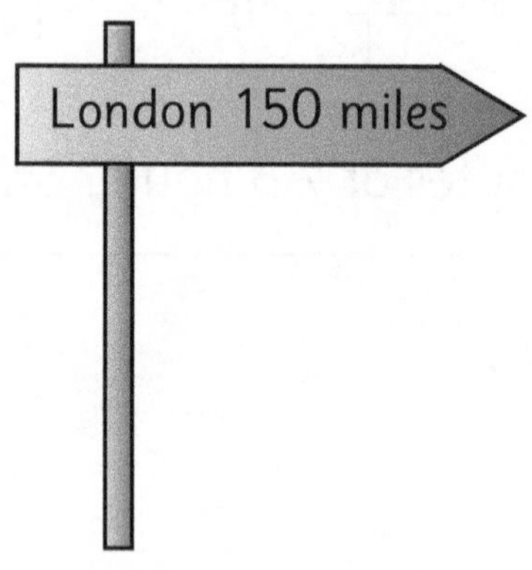

km

4. Which is the faster speed, 96 km per hour or 65 miles per hour? Show how you know.

Unit 35: Convert between miles and kilometres

1. Convert kilometres to miles, using 5 miles = 8 kilometres.

a) 45 miles = ☐ km

b) 120 miles = ☐ km

2. Convert the miles to kilometres, using 1 mile = 1·6 kilometres.

a) 4 miles = ☐ km

b) 12 miles = ☐ km

3. Using the line graph to help you, show whether 70 miles per hour or 120 kilometres per hour is the faster speed. Show your working.

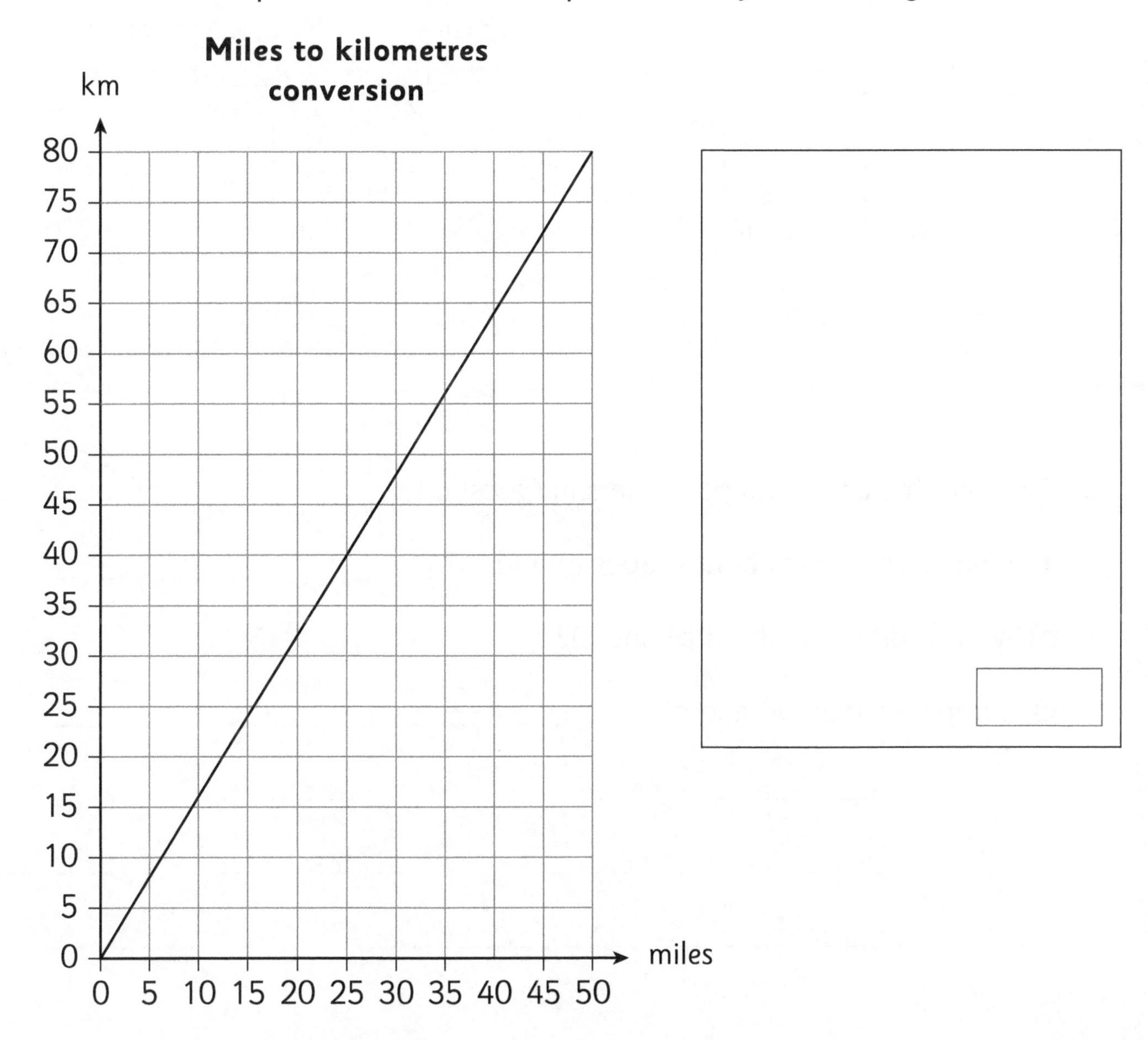

Unit 36: Recognise that shapes with the same area can have different perimeters and vice versa

1. Find the area and perimeter of each shape.

a)

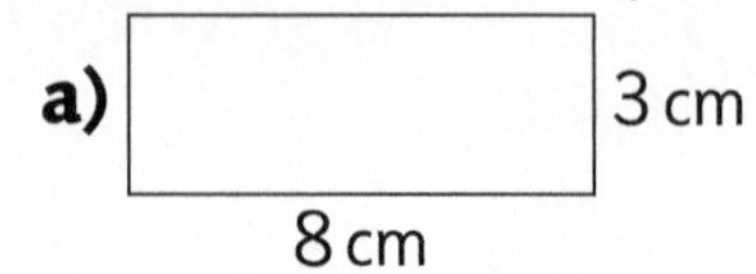

A = ☐ cm^2

P = ☐ cm

b)

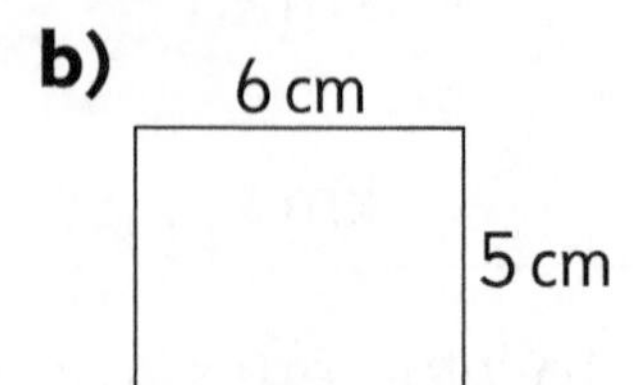

A = ☐ cm^2

P = ☐ cm

c)

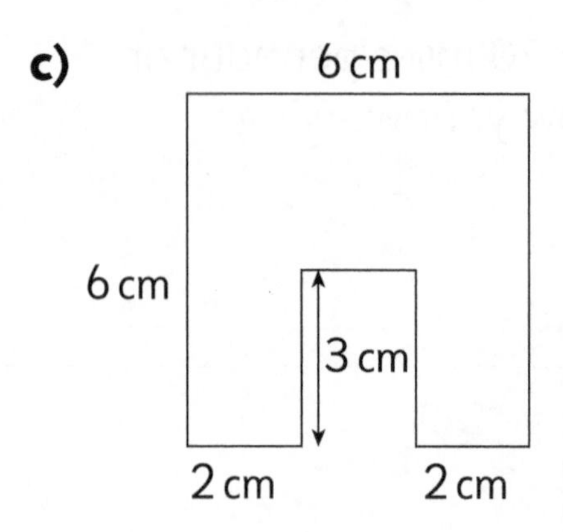

A = ☐ cm^2

P = ☐ cm

d)

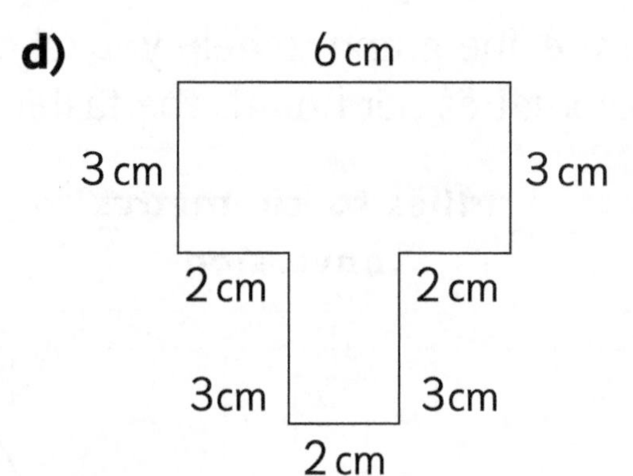

A = ☐ cm^2

P = ☐ cm

2. Compare the areas and perimeters in Question 1.

a) What is the same about shapes a) and b)? ________________________

b) What is different about b) and c)? ________________________

c) Compare shapes a) and d). ________________________

____

____

____

Unit 36: Recognise that shapes with the same area can have different perimeters and vice versa

1. Show on the squared grid that two rectangles with the **same area** do not have the same perimeters.

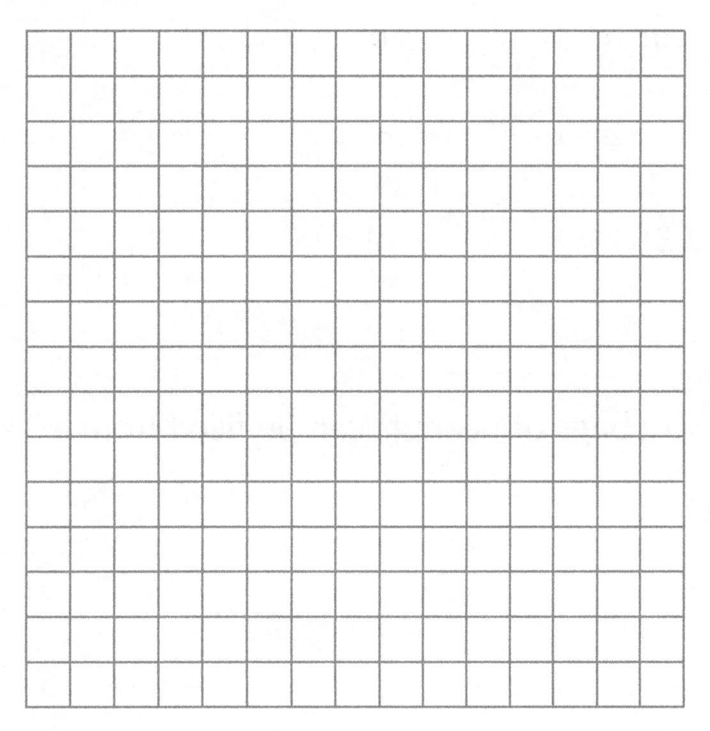

2. Compare the two shapes that are similar, referring to their area or perimeter.

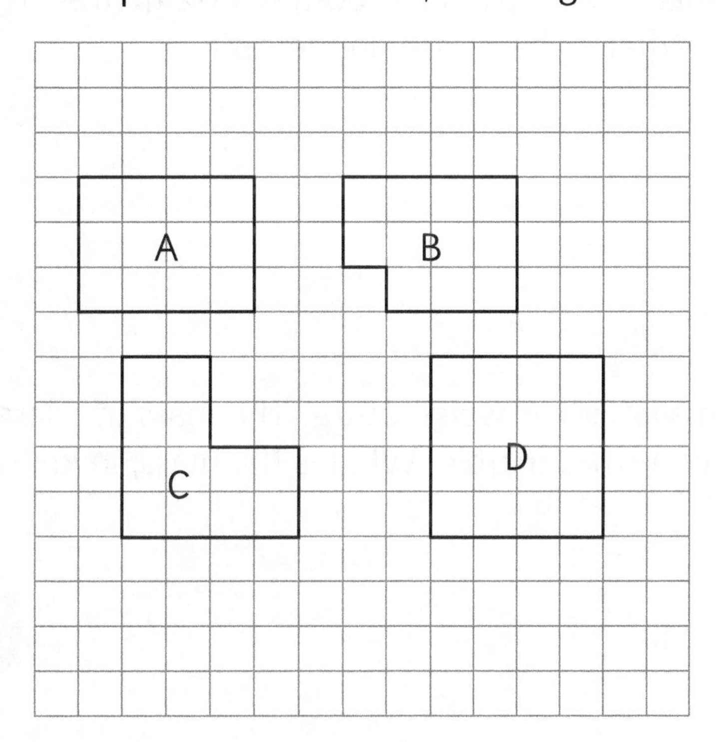

What is the same? What is different?

____

Unit 37: Solve problems involving the conversion of units of measure

1. Solve these calculations, giving your answers in **both** units of measure.

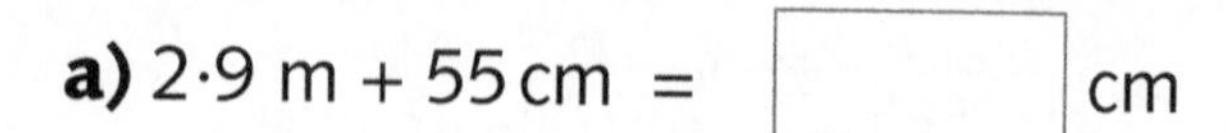

a) 2·9 m + 55 cm = ☐ cm

= ☐ m

b) 7200 ml – 2·36 l = ☐ ml

= ☐ l

2. How many 250 g bags of peanuts can be filled from a 5 kg sack?

☐ bags

3. The total perimeter of a rectangular room is 11·2 metres. The length of the room is 320 cm. What is the length, in metres?

☐ m

4. One hundred small marbles weigh 256 g. The mass of a large marble is three times the mass of a small marble. What is the mass, in kg, of 500 large marbles?

☐ kg

Unit 37: Solve problems involving the conversion of units of measure

1. A wall is built of bricks, each with a depth of 8 cm. The wall is 3·44 metres high. How many bricks tall is the wall?

   ☐ bricks

2. Tom divides 4·5 kg of potatoes into 3 bags. One bag has a mass of 2·56 kg, the second bag contains 750 g. What is the mass of the third bag?

   ☐ kg

3. After a school fair, the stall holder is counting the money. He has sixteen 20p coins, eighty-two 10p coins and £7·83 in pennies. How much is it altogether?

   £ ☐

4. Ayesha is running a long distance race of 40 km. She runs for 2·1 km, then sprints for 400 m, runs for 2·1 km, sprints for 400 m. She keeps repeating this pattern. How many sets of runs/sprints does she need to do to complete the race?

   ☐ sets

Unit 38: Calculate the area of parallelograms and triangles

1.

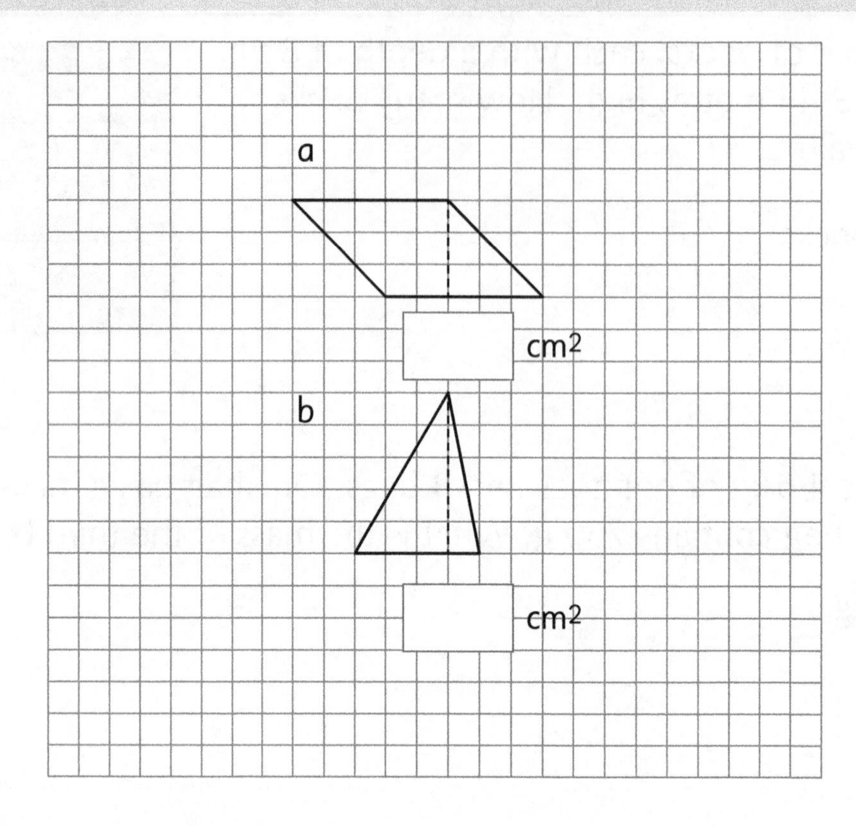

a) Find the area of the parallelogram.

b) Find the area of the triangle.

2. Find the areas of each shape.

a)

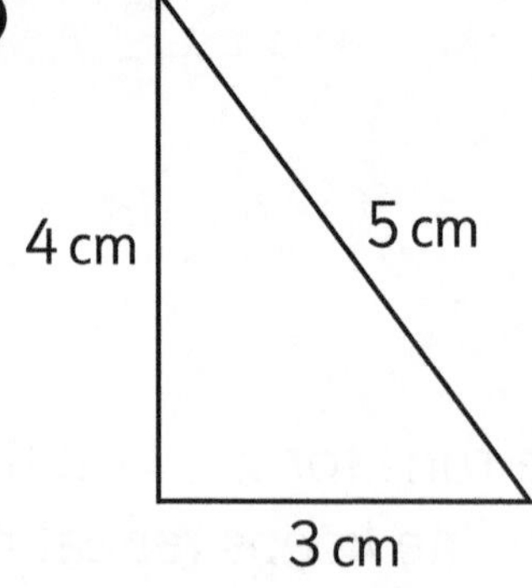

cm²

b)

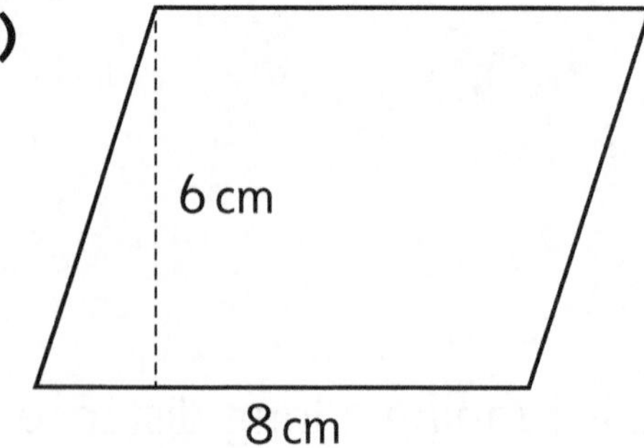

cm²

Unit 38: Calculate the area of parallelograms and triangles

1.

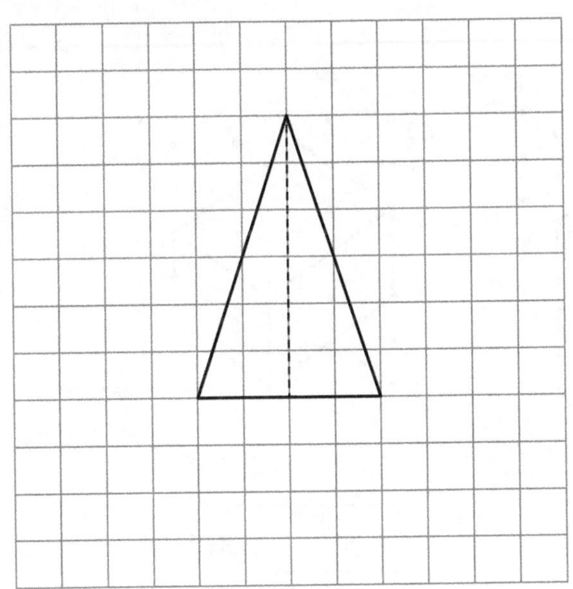

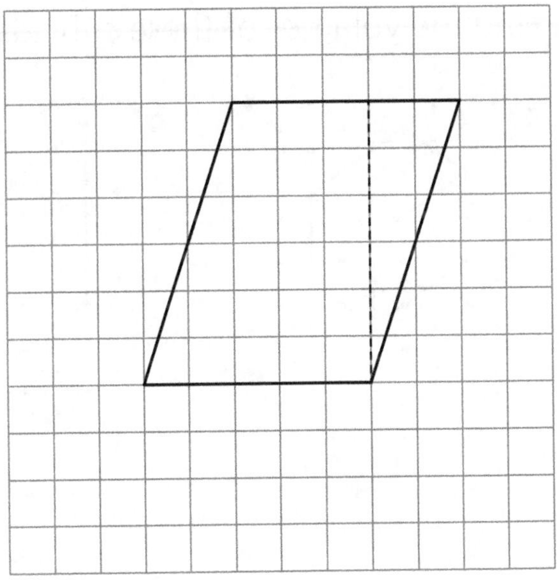

a) Find the area of the triangle.  cm^2

b) Find the area of the parallelogram.  cm^2

2. The area of the rectangle shown in the dotted lines is 28 cm^2.

a)

b)

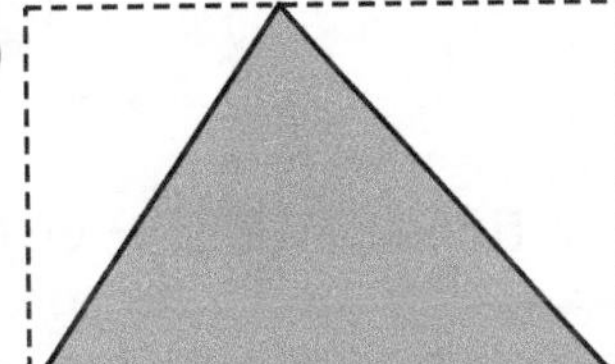

a) What is the area of the parallelogram?  cm^2

b) What is the area of the triangle? 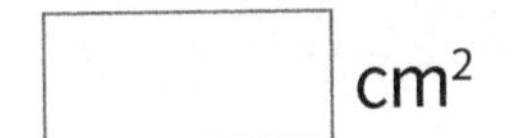 cm^2

Unit 39: Calculate, estimate and compare the volume of cubes and cuboids

1. Find the volumes of these cuboids.

a)

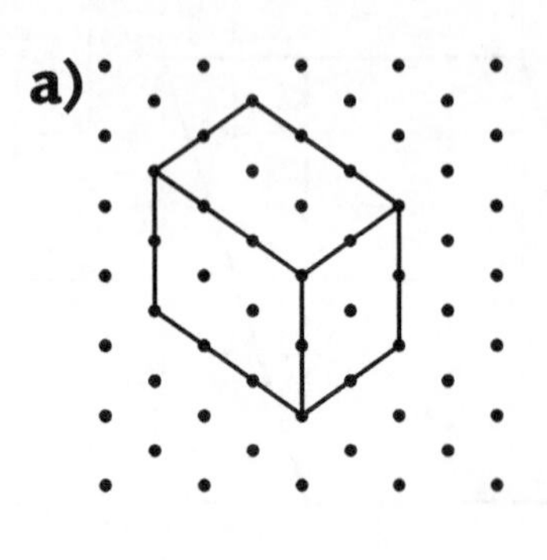

b)

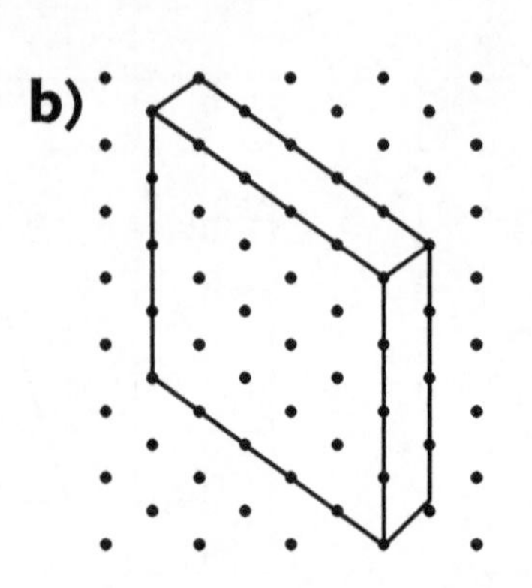

c)

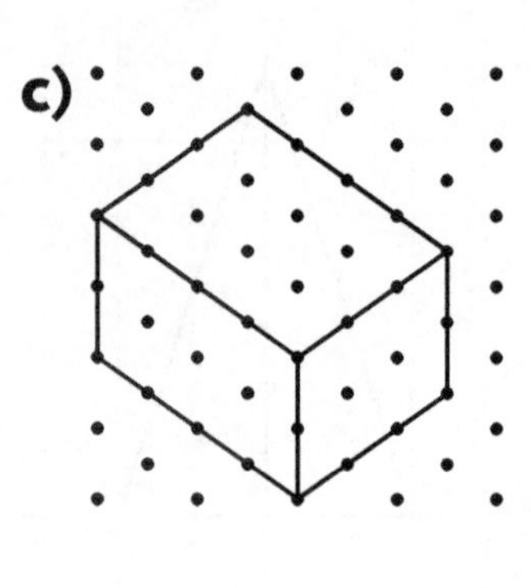

☐ cm^3 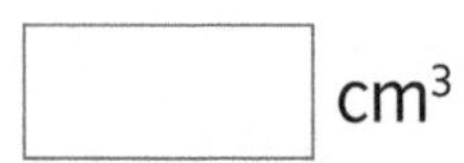 cm^3  cm^3

2. The volume of an art sculpture in the form of a cube is 64 m^3. How tall is it?

☐ m

3. Write these volumes using the correct cubic units.

a) 4 mm × 7 mm × 5 mm = ☐

b) 4·5 cm × 2 cm × 6 cm = ☐

4. A cuboid has a volume of 30 cm^3. What could its dimensions be?

length ☐ cm width ☐ cm height ☐ cm

Unit 39: Calculate, estimate and compare the volume of cubes and cuboids

1. Which shape has a **different** volume?

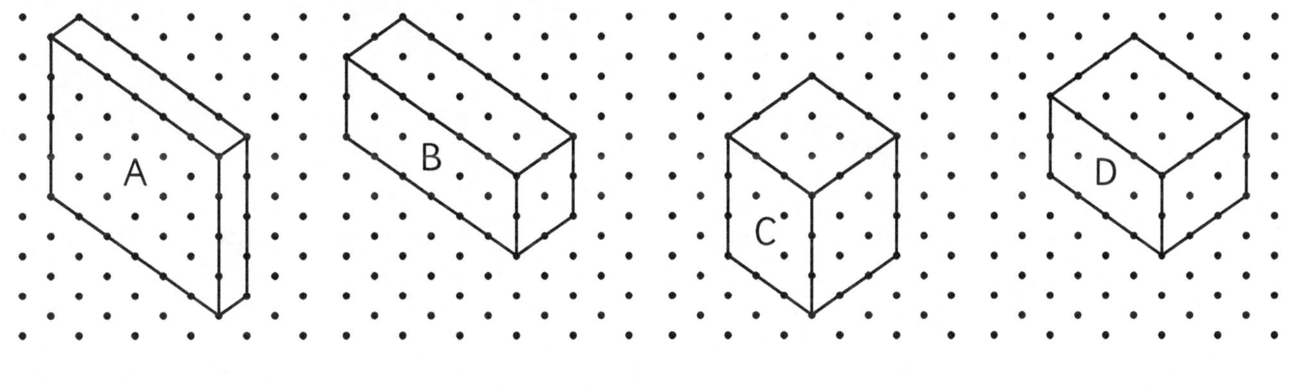

Shape ________

2. How many cubic metres larger is A than B? Show how you know.

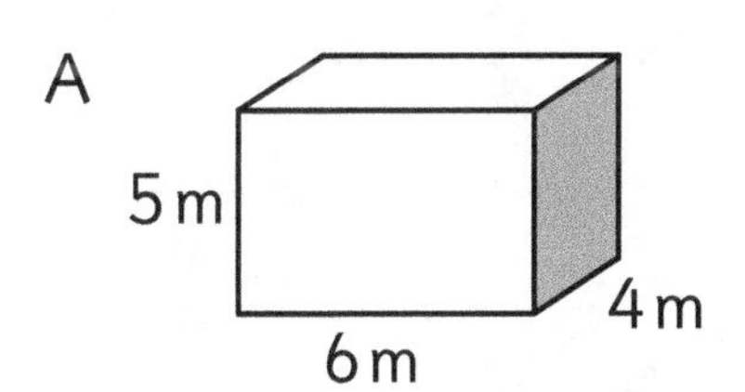

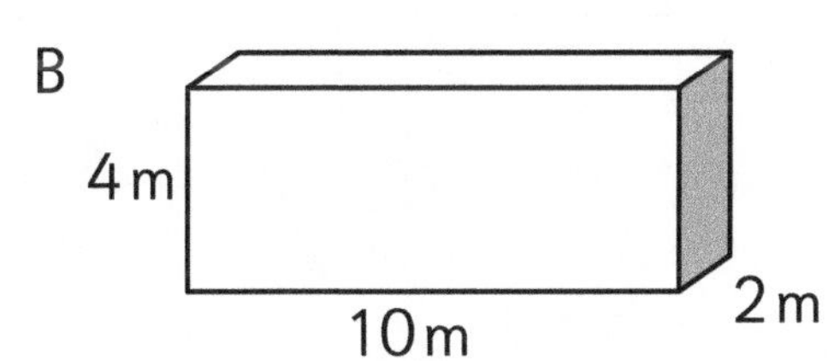

3. These toy bricks are cubes 2 cm long.
How many will fit into a box measuring
8 cm by 6 cm by 4 cm? [ ]

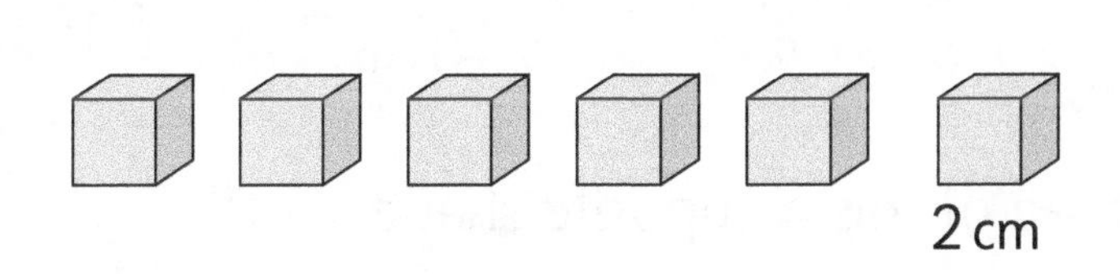

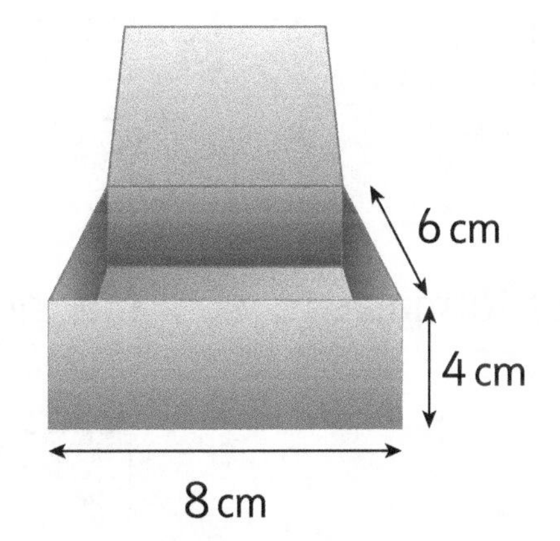

Unit 40: Recognise when it is possible to use formulae for the area and volume of shapes

1. Draw lines to connect the formulae to the correct shapes.

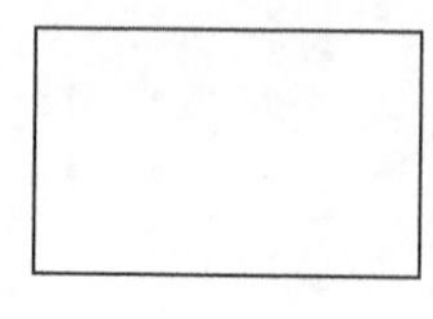

$V = w^3$

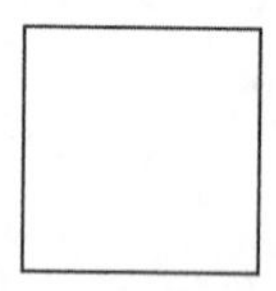

$A = \frac{1}{2}bh$

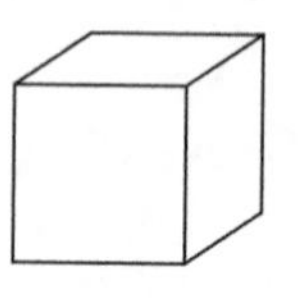

$A = wl$

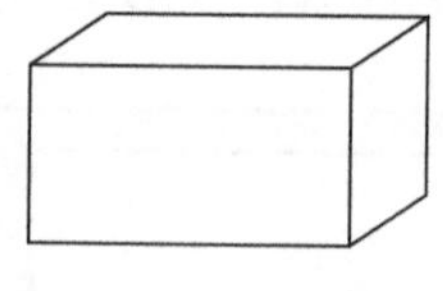

$A = bh$

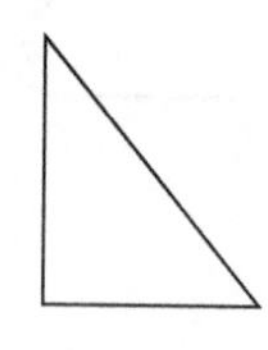

$A = w^2$

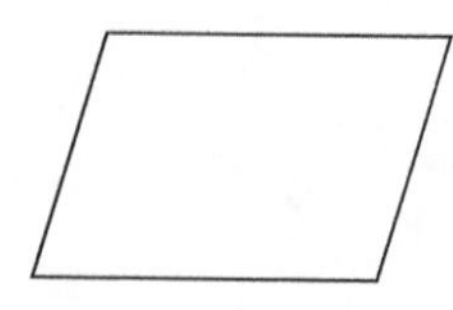

$V = lwh$

2. Tick the shapes that show the correct dimensions labelled to find the area or volume.

a) **b)**

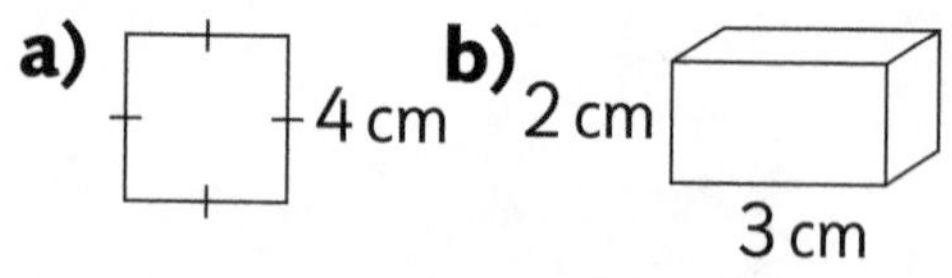

c)

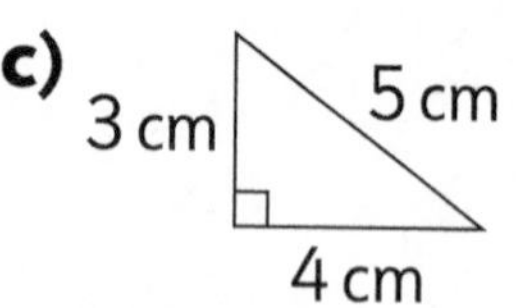

d)

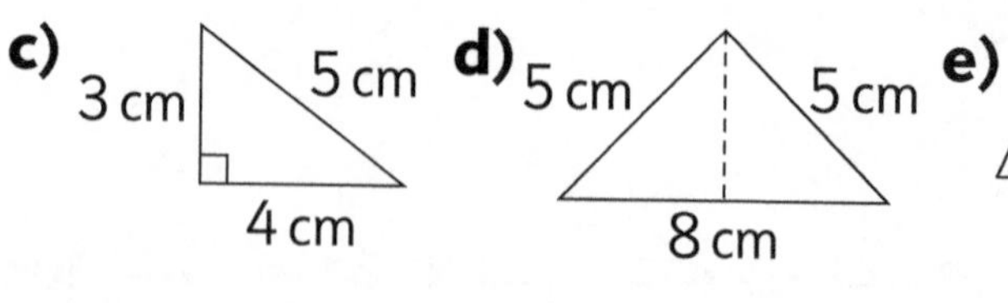

e)

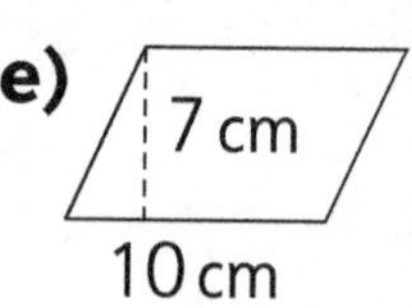

3. Use formulae to work out the area of this composite shape.

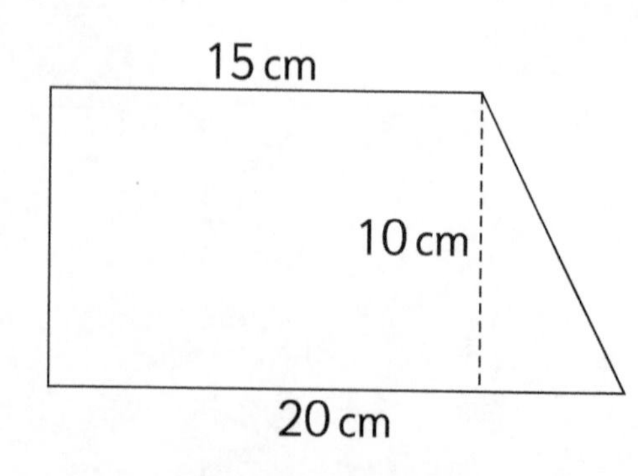

Area = ☐ cm^2

Unit 40: Recognise when it is possible to use formulae for the area and volume of shapes

1. Which dimensions are needed to find the following areas or volumes?

a) area of a triangle ________ ________

b) volume of a cube ________

c) area of a parallelogram ________ ________

d) volume of a cuboid ________ ________ ________

2. Split the 2-D shape to work out its area.

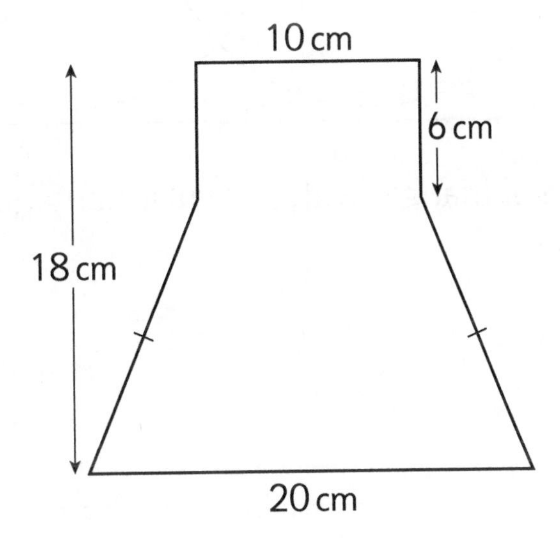

Area = ☐ cm²

3. Find the area of this trapezium when a = 8 cm, b = 10 cm and h = 6 cm.

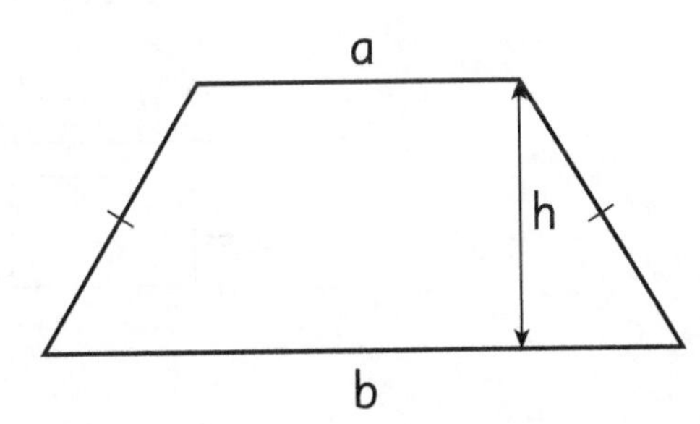

Area = ☐ cm²

Unit 41: Compare and classify geometric shapes based on their properties and sizes and find unknown angles in any triangles, quadrilaterals, and regular polygons

1. Write the names of the shapes in the correct places on the Venn diagram, according to their properties.

square

rhombus

parallelogram

kite

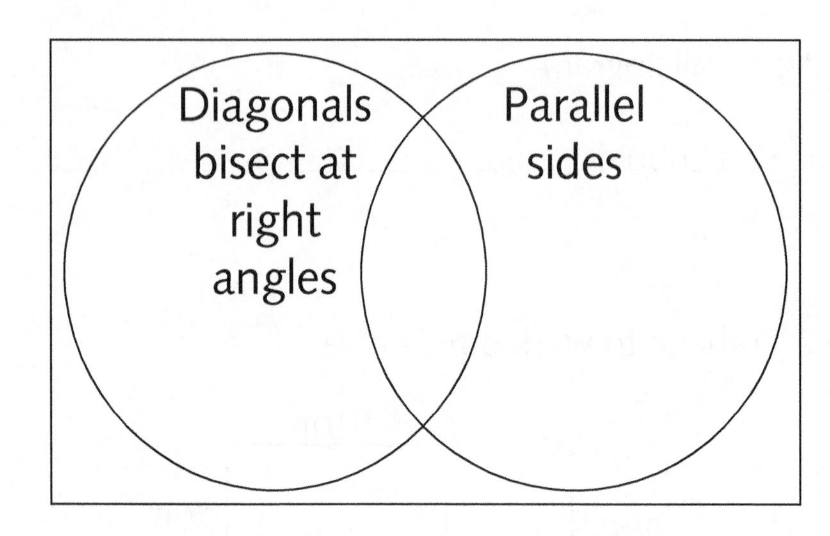

2. Compare an equilateral triangle with a regular pentagon.

a) What is the same?

____

b) What is different?

____

3. Find the missing angles in the triangles.

a)

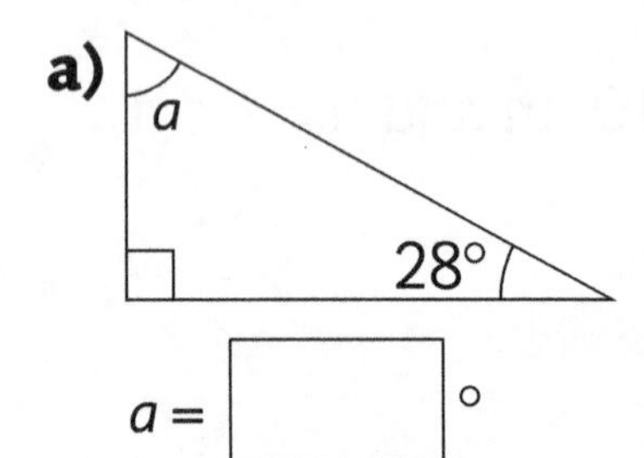

b)

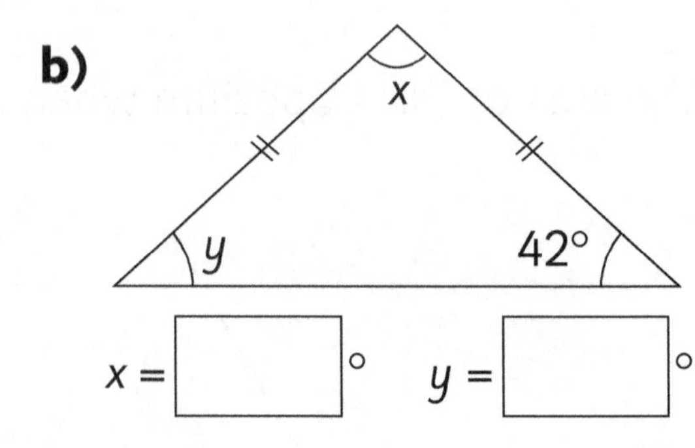

Unit 41: Compare and classify geometric shapes based on their properties and sizes and find unknown angles in any triangles, quadrilaterals, and regular polygons

4. Find the missing angles marked with a letter.

a)

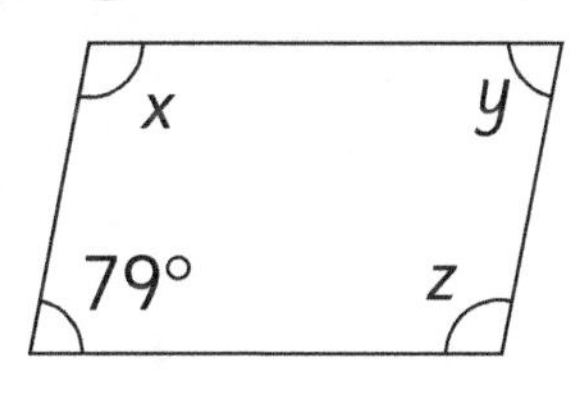

Not drawn to scale

x = ☐ ° y = ☐ ° z = ☐ °

b)

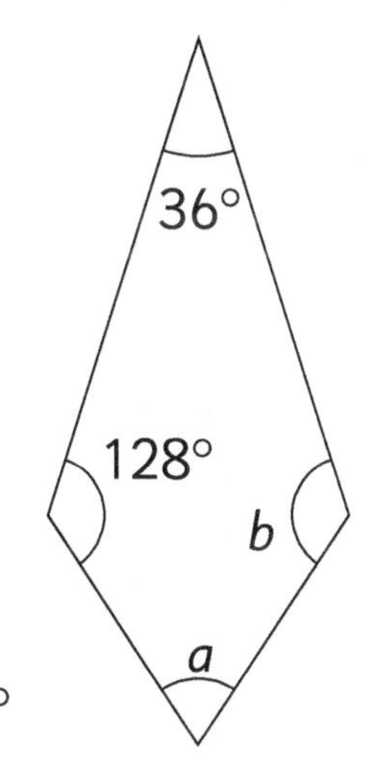

Not drawn to scale

a = ☐ ° b = ☐ °

5. Work out the size of angle p in this quadrilateral.

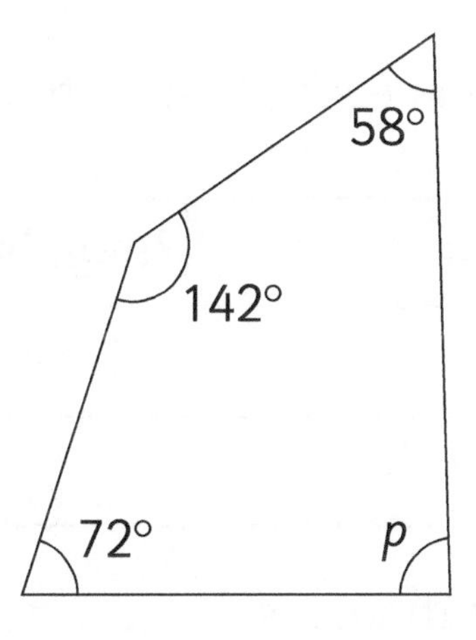

Not drawn to scale

p = ☐ °

Unit 41: Compare and classify geometric shapes based on their properties and sizes and find unknown angles in any triangles, quadrilaterals, and regular polygons

1.

A

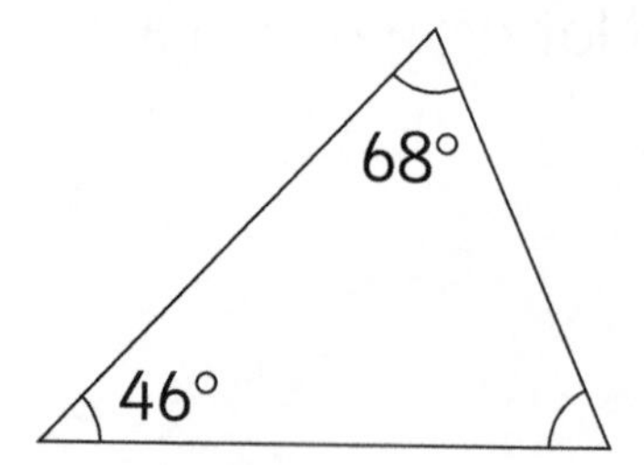

B

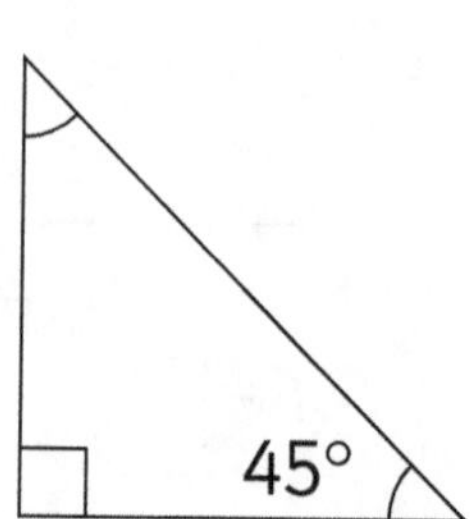

C

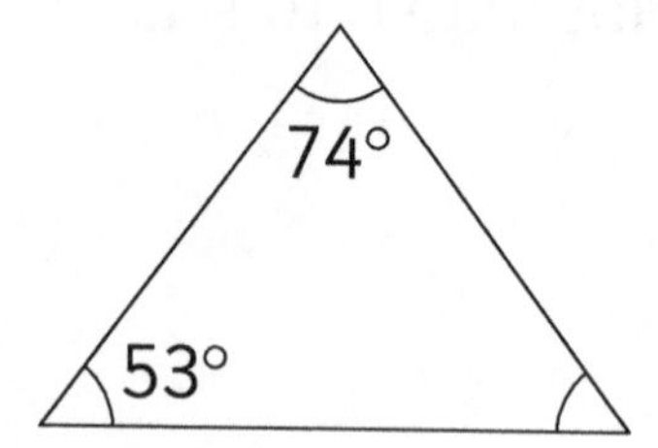

D

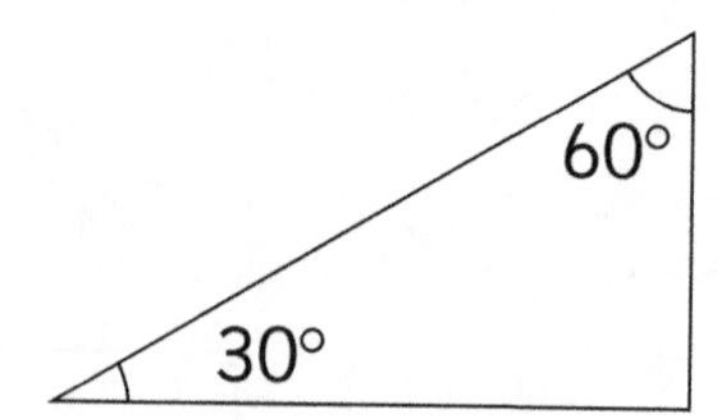

a) Write the letters of the isosceles triangles. __________

b) Write the letters of the right-angled triangles. __________

2. Write the name of a shape in each place on the Carroll diagram.

	regular	**irregular**
triangle		
quadrilateral		

3. Write three properties of a rhombus.

__________ __________ __________

Unit 41: Compare and classify geometric shapes based on their properties and sizes and find unknown angles in any triangles, quadrilaterals, and regular polygons

4. Find the angle marked *y* on this trapezium.

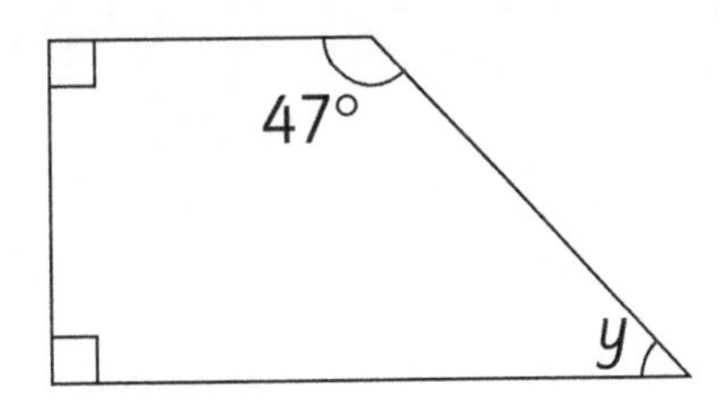

$y =$ ☐ °

5. Lenny has measured the angles in the quadrilateral. Has he measured them correctly? Show how you know.

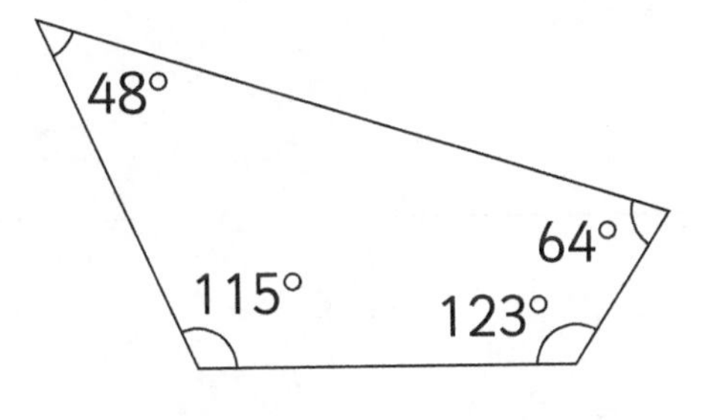

____

____

6. Label the kite to show the equal sides, equal angles and right angles.

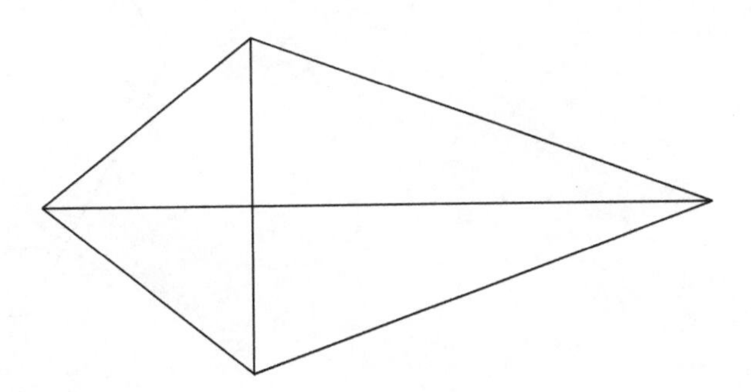

Unit 42: Draw 2-D shapes using given dimensions and angles

1. The triangle is not drawn to scale. Draw the full-scale triangle accurately. The base has been drawn for you.

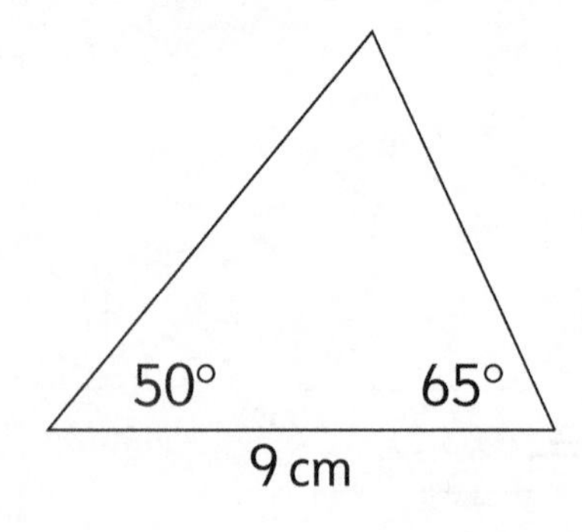

2. The inside angles of a regular hexagon are 120°. Complete the drawing of a regular hexagon with sides of length 5 cm.

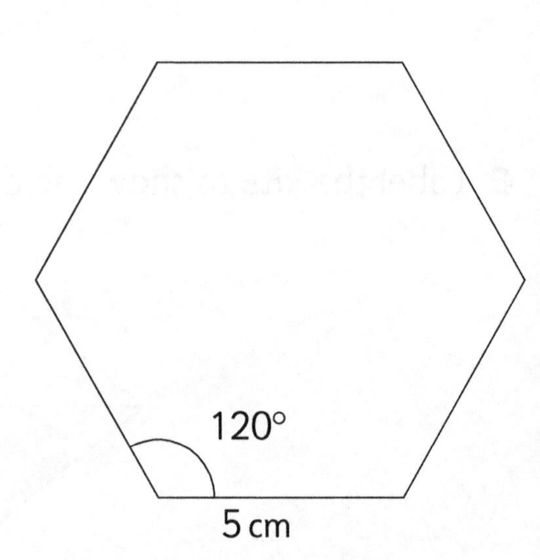

Unit 42: Draw 2-D shapes using given dimensions and angles

1. The triangle is not drawn to scale. Draw the full-scale triangle accurately. The base has been drawn for you.

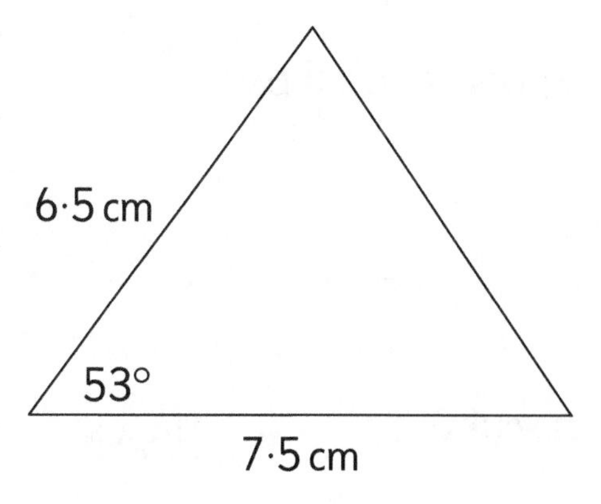

2. The kite is not drawn to scale. Complete the full-scale kite accurately.

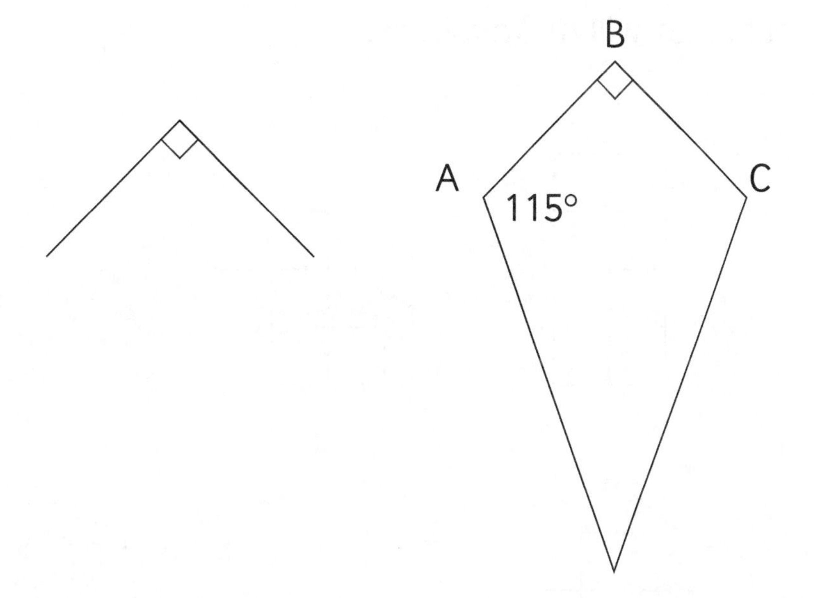

Unit 43: Recognise, describe and build simple 3-D shapes, including making nets

1. How many faces do each of these 3-D shapes have?

a) pentagonal prism ☐ faces

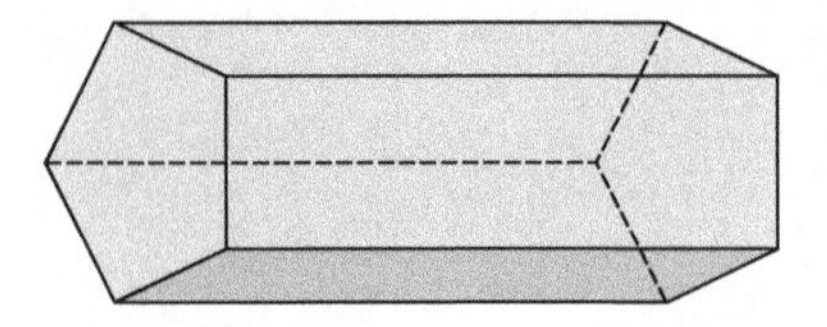

b) cuboid ☐ faces

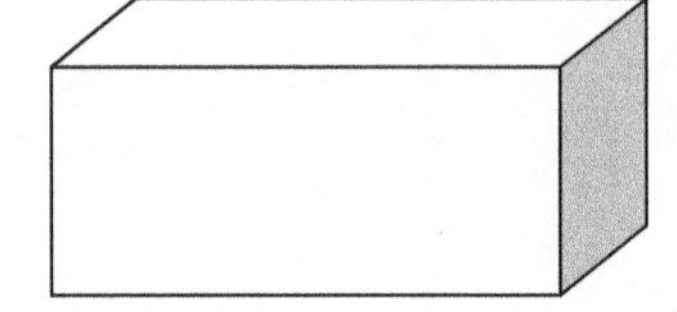

c) square-based pyramid ☐ faces

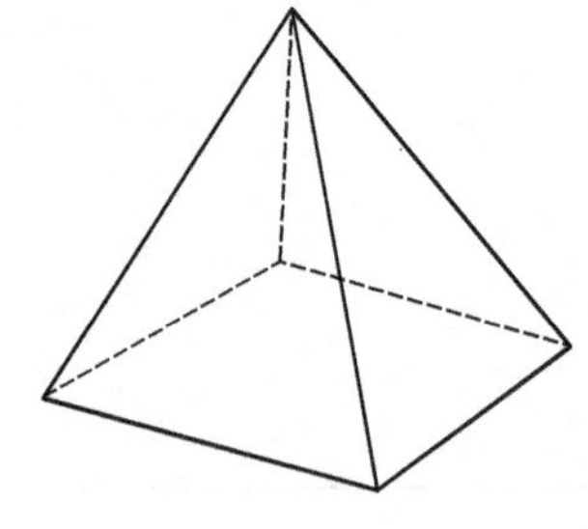

2. Circle all the nets that will make a cube.

A

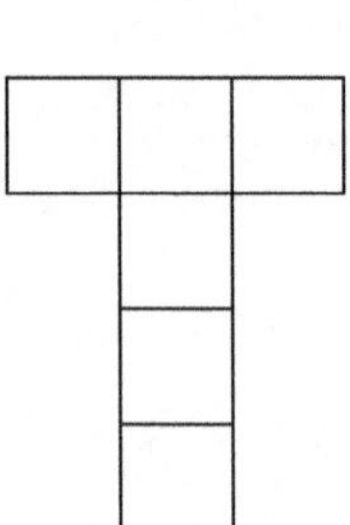

B

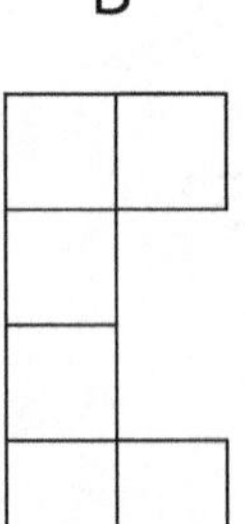

C

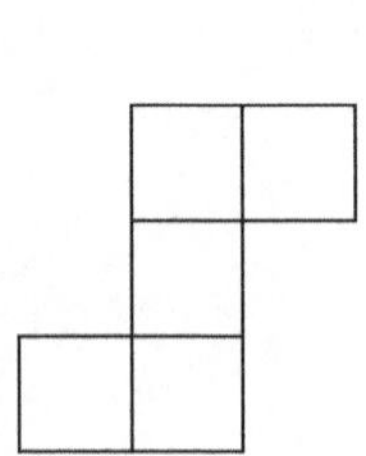

D

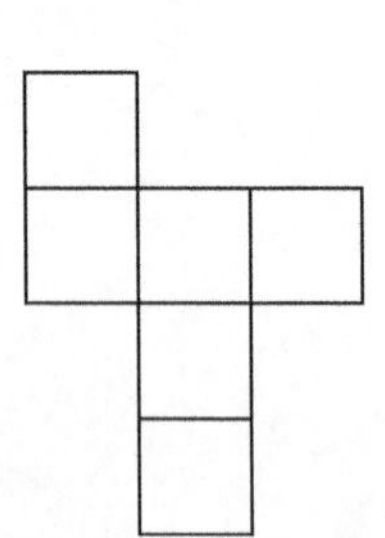

E

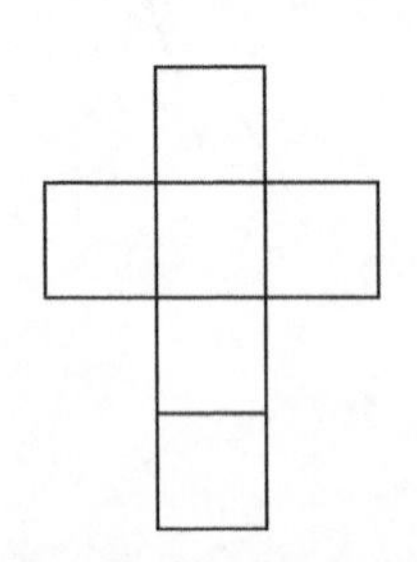

Unit 43: Recognise, describe and build simple 3-D shapes, including making nets

3. Draw the net of a triangular prism.

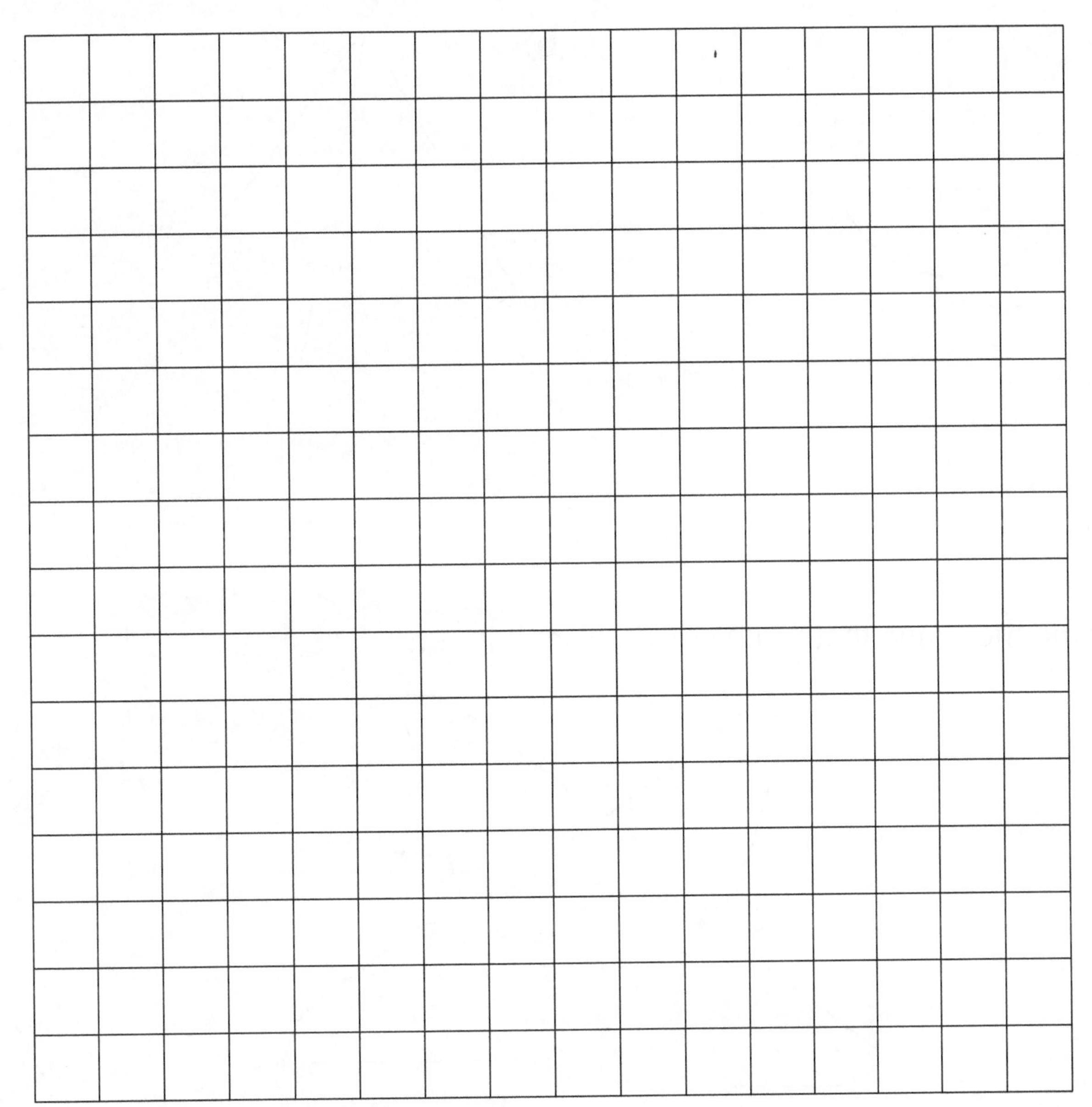

4. This is the net of a 3-D shape. What is the name of the shape?

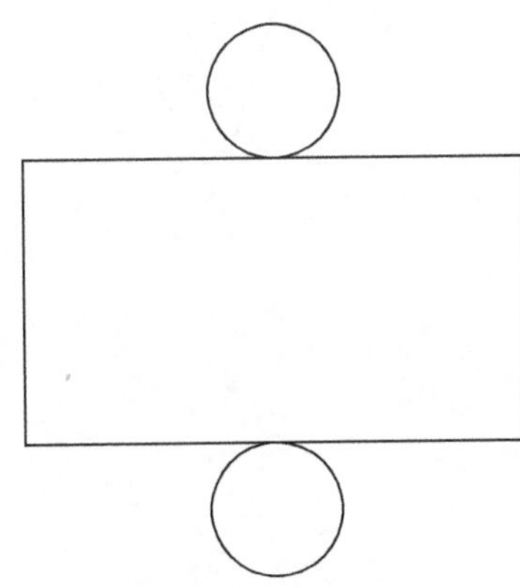

Unit 43: Recognise, describe and build simple 3-D shapes, including making nets

1. Name the shapes.

a)

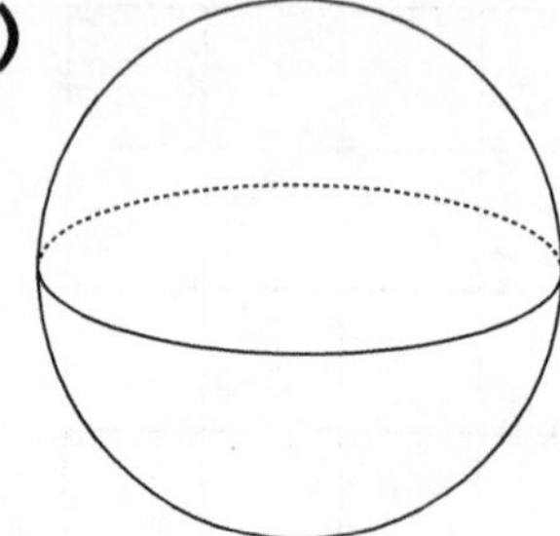

b)

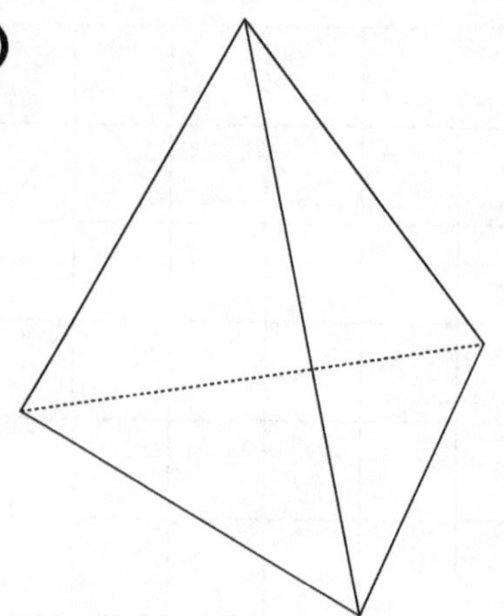

____________________ ____________________

2. Name the shape that will be made from this net.

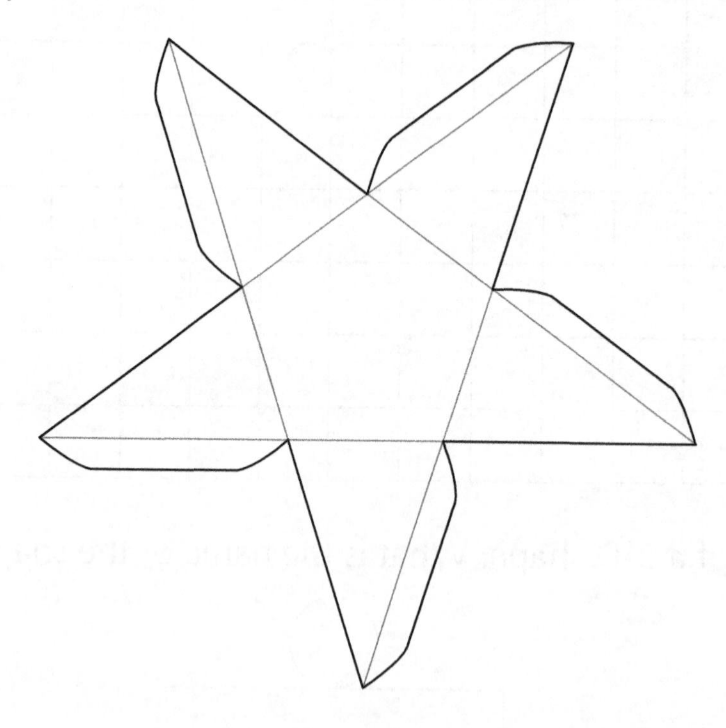

____

Unit 43: Recognise, describe and build simple 3-D shapes, including making nets

3. Complete the net of this cuboid.

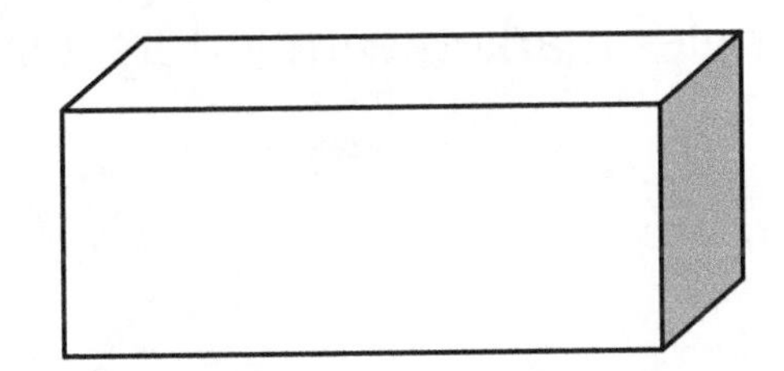

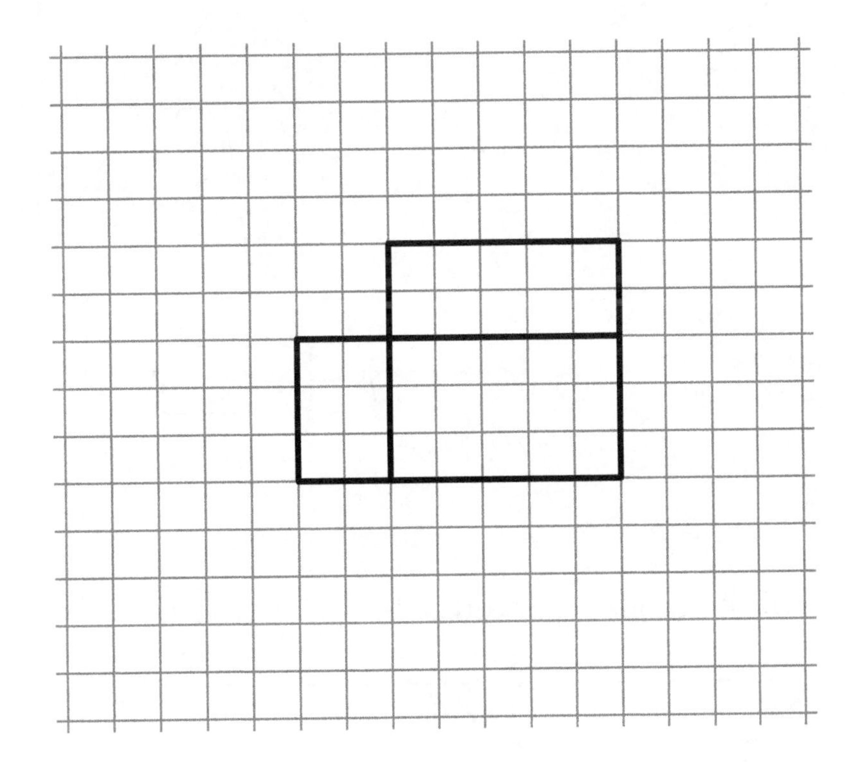

4.

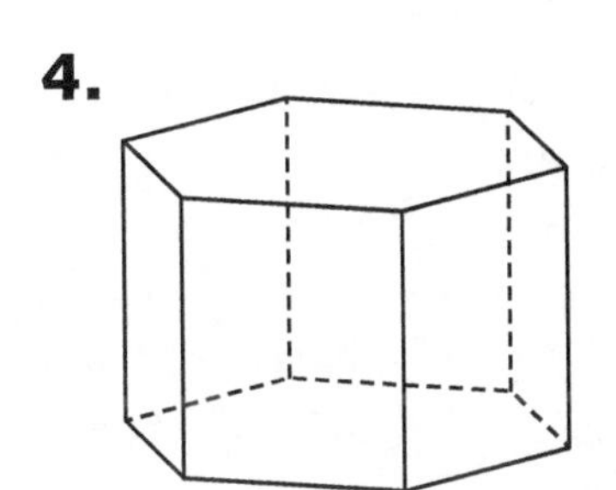

a) How many faces does a hexagonal prism have?

_____________ faces

b) How many edges does a hexagonal prism have?

_____________ edges

Unit 44: Recognise angles where they meet at point, are on a straight line, or are vertically opposite, and find missing angles

1. Find the value of the angles marked with letters, using angle properties.

a)

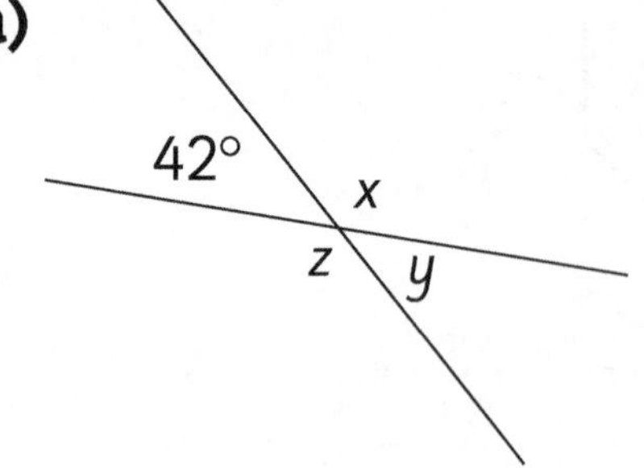

b)

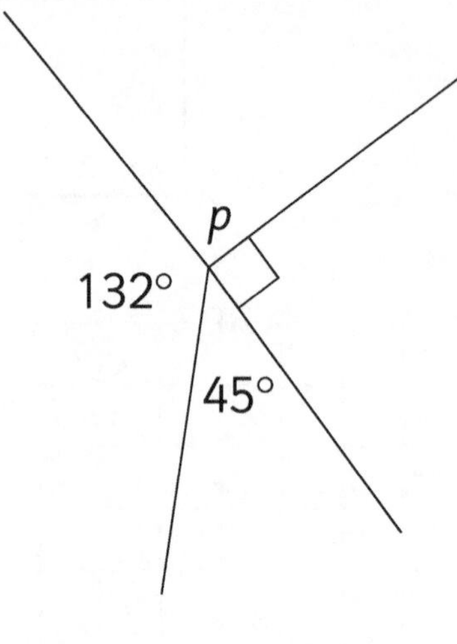

x = [ ]°

y = [ ]°

z = [ ]°

p = [ ]°

2. Find the value of the angles marked *a*, *b*, *c*, *d*, *e* and *f*.

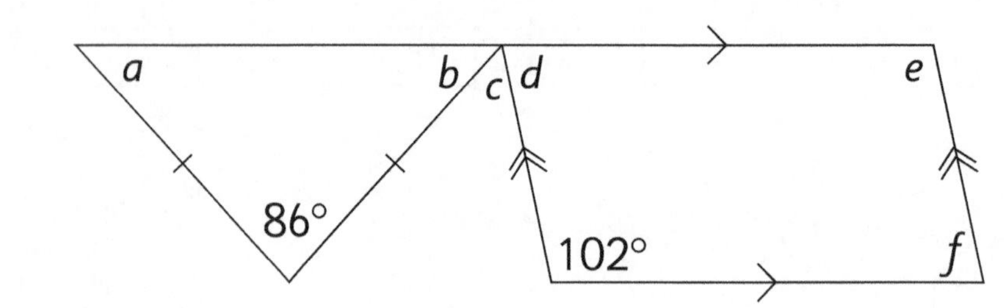

a = [ ]° b = [ ]° c = [ ]°

d = [ ]° e = [ ]° f = [ ]°

Unit 44: Recognise angles where they meet at point, are on a straight line, or are vertically opposite, and find missing angles

1. Find the value of the angle marked y.

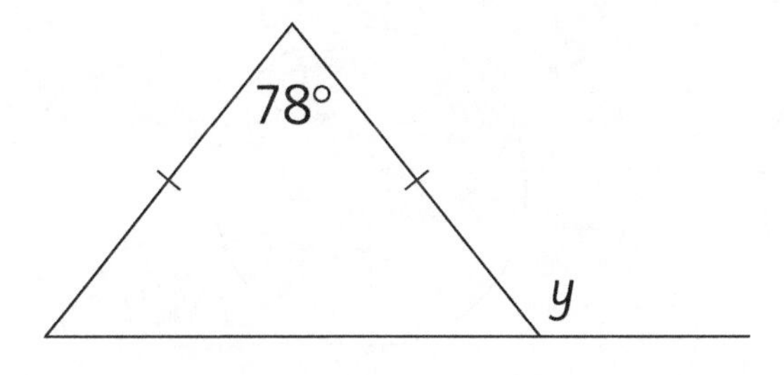

$y =$ ☐ °

2. Work out the value of the angles marked p and q.

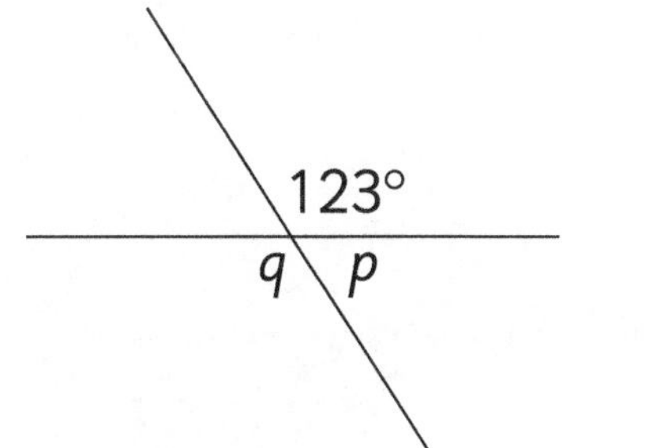

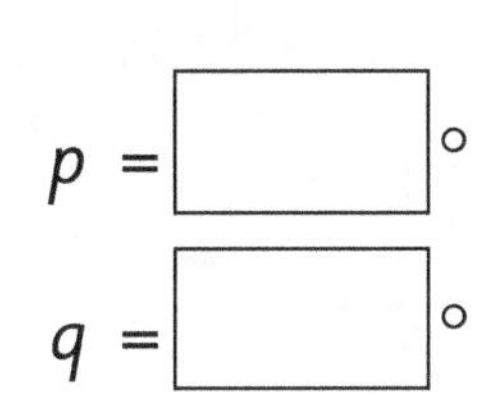

3. Priya says that the angle marked x is 60°. Prove that she is right using the angle sums of triangles, quadrilaterals and straight lines.

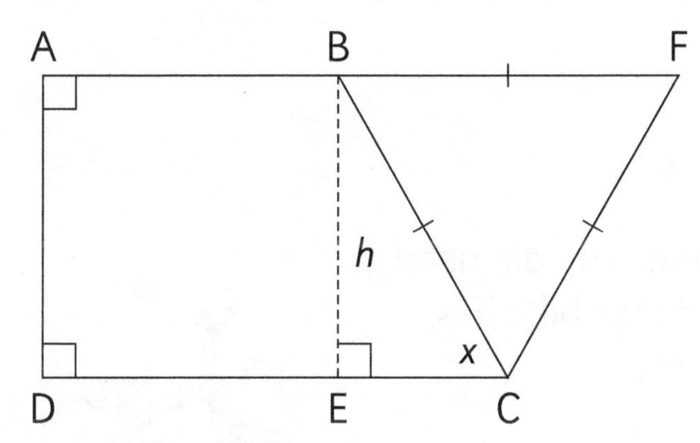

____

____

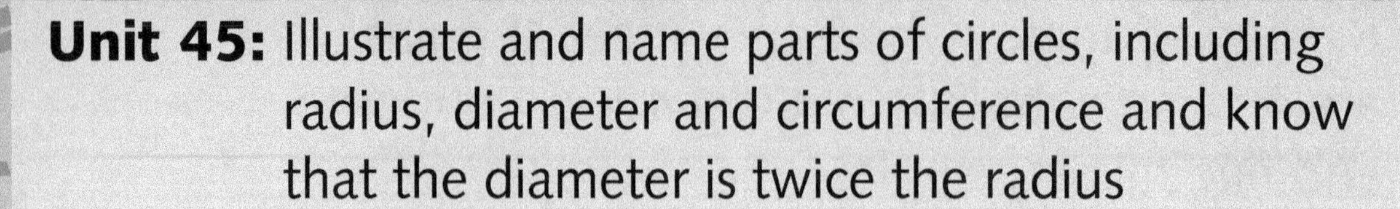

Unit 45: Illustrate and name parts of circles, including radius, diameter and circumference and know that the diameter is twice the radius

1. Use a coloured pencil or pen to show these parts of a circle.

a)
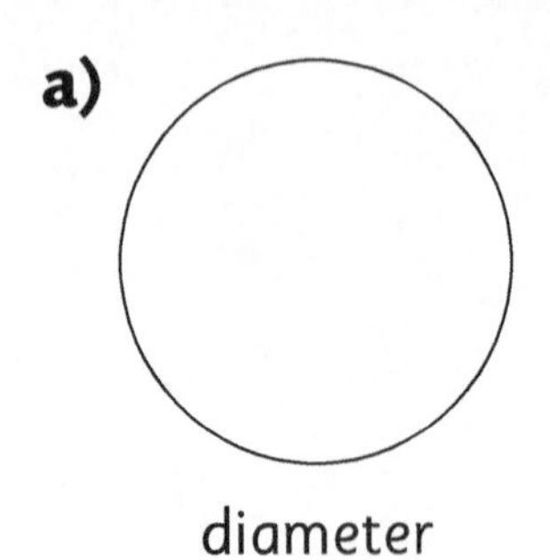
diameter

b)
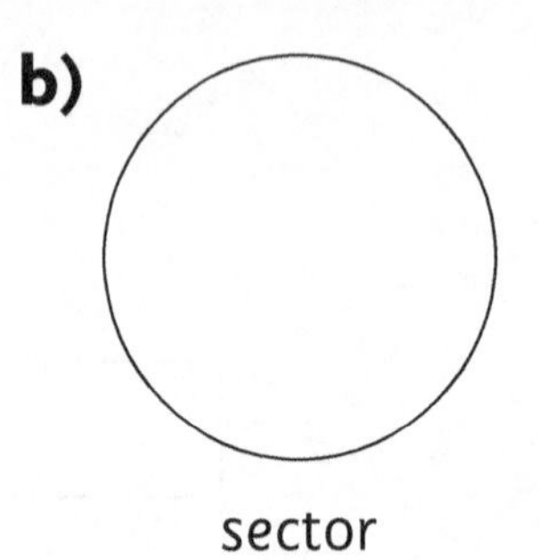
sector

c)
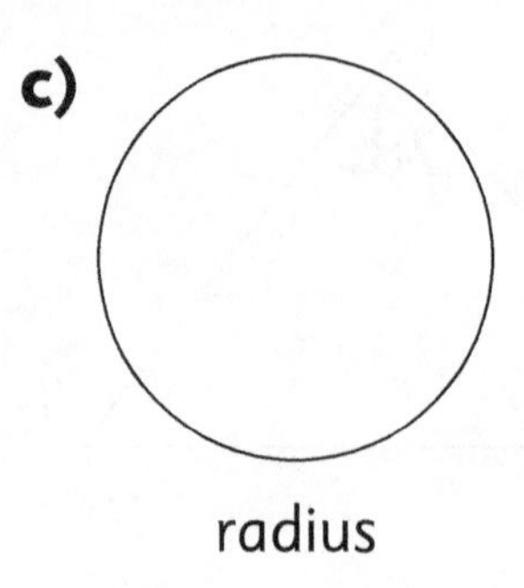
radius

d)
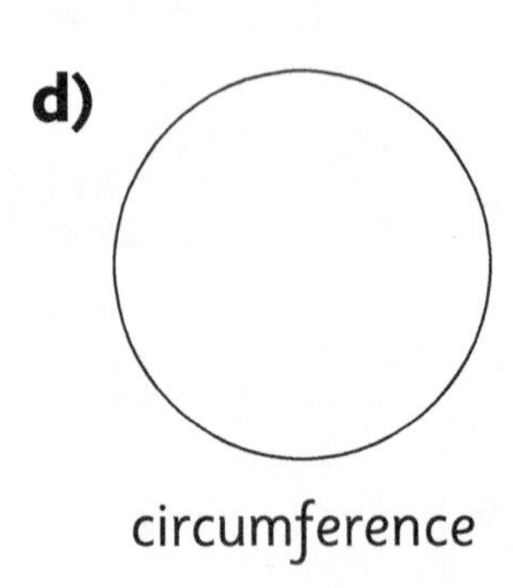
circumference

e)
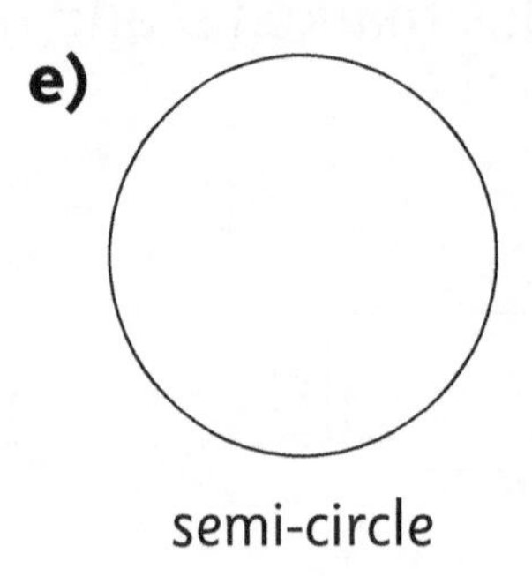
semi-circle

2. Write the formula to show the connection between the radius and the diameter of a circle. Use 'r' for radius and 'd' for diameter.

____

3. Ollie's bike wheel has a radius of 15 cm. The diameter of the wheel on Ali's bike is 28 cm. Whose bike has the larger wheel? Show how you know.

Unit 45: Illustrate and name parts of circles, including radius, diameter and circumference and know that the diameter is twice the radius

1. Write the part of the circle shown in bold.

a)

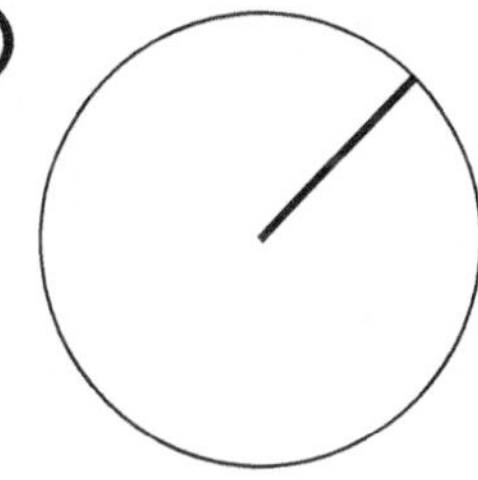

____

b)

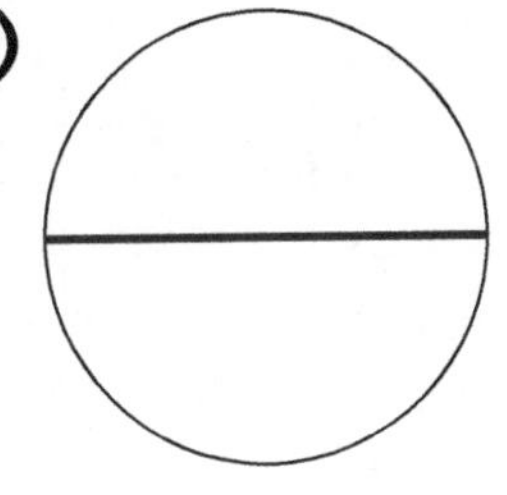

____

c)

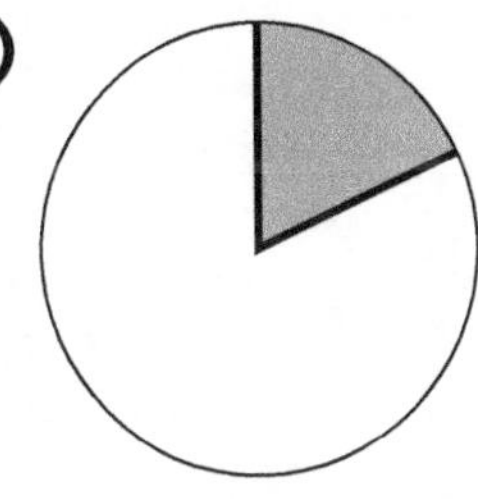

____

d)

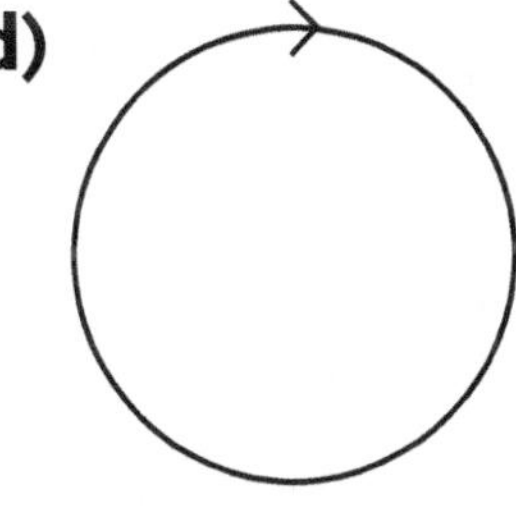

____

e)

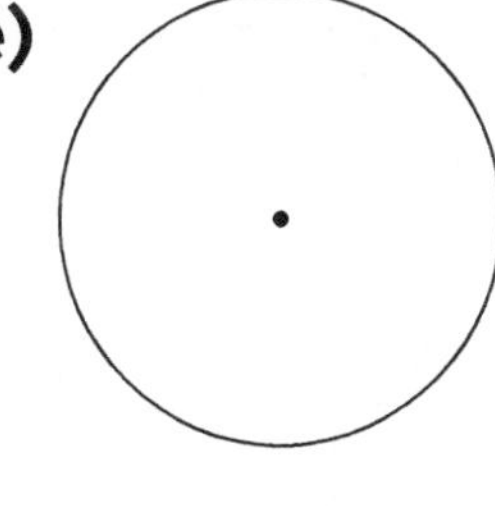

____

f)

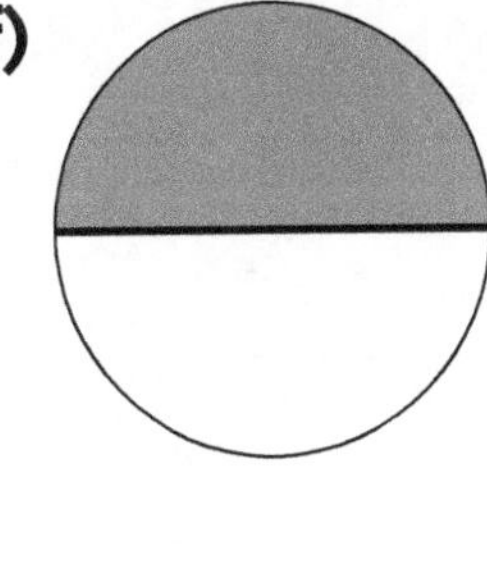

____

2. The radius of a circle is 7·6 cm. What is its diameter?

______________ cm

3. The approximate circumference of a circle can be found using the formula:

circumference = 3·142 × diameter

Find the circumference of each circle.

a)

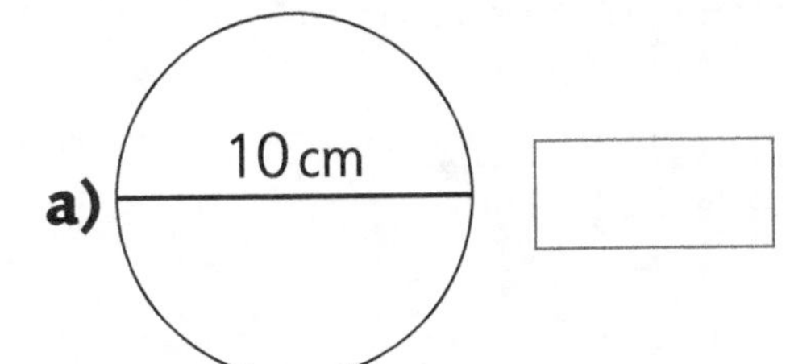

b)

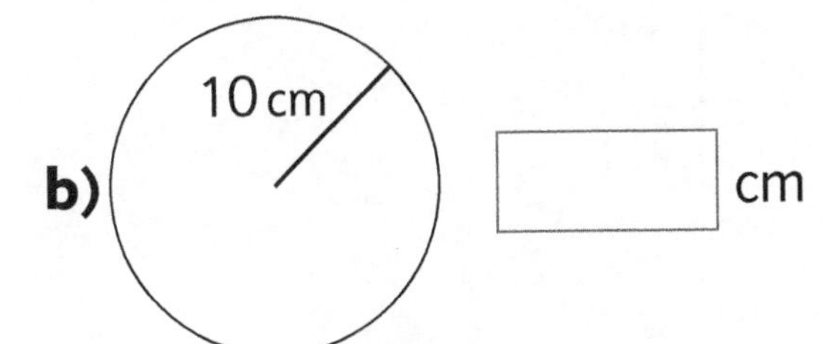

Unit 46: Describe positions on the full coordinate grid

1. Plot these coordinates on the grid, labelling them A, B, C, D, E and F.

A (7, 0) B (0, 4) C (–3, –5) D (–6, 1) E (1, –6) F (0, 0)

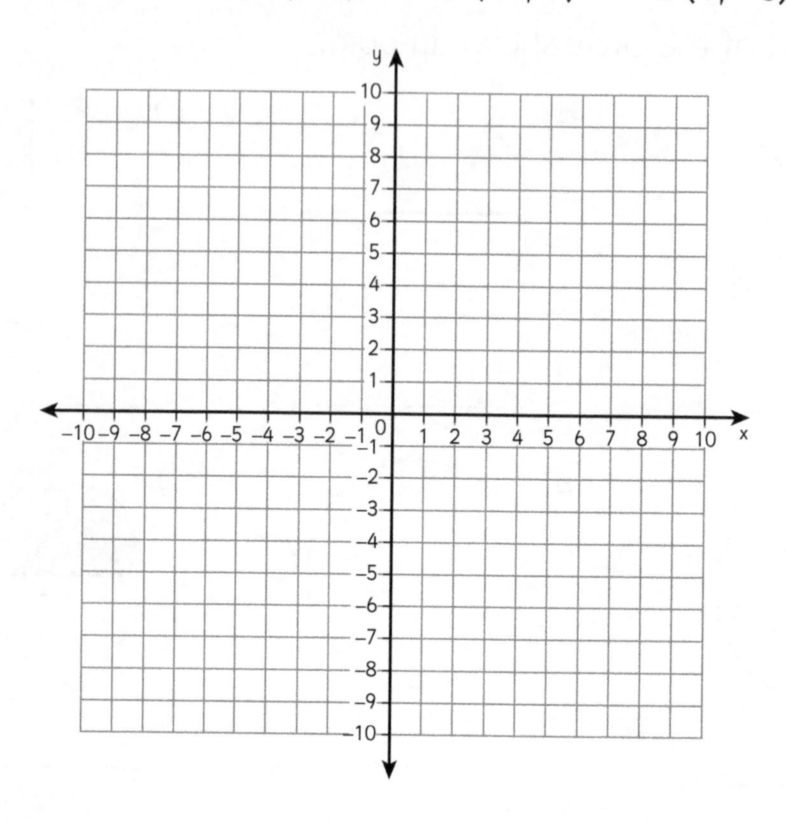

2. A rectangle has vertices (–4, 1) and (–2, 3). What are the coordinates of the other two vertices? Draw the rectangle and label the vertices.

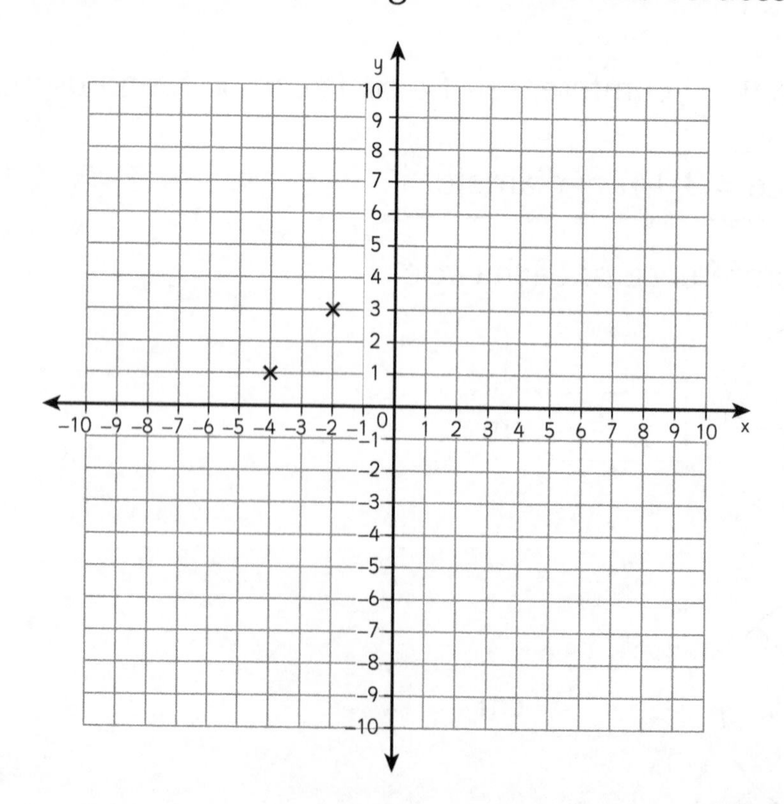

Unit 46: Describe positions on the full coordinate grid

1.

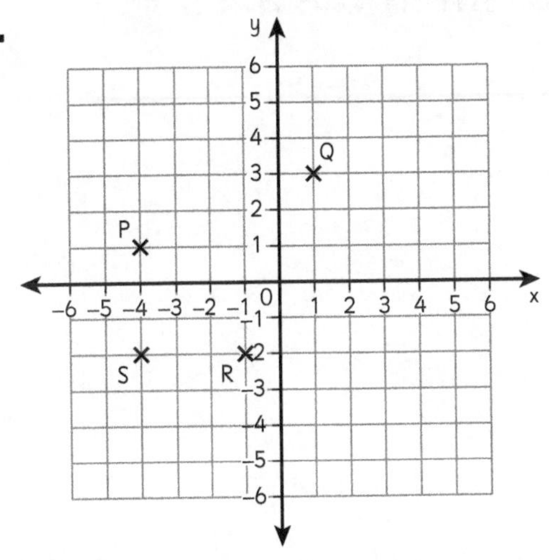

a) Write the coordinates of these points.

P (,)

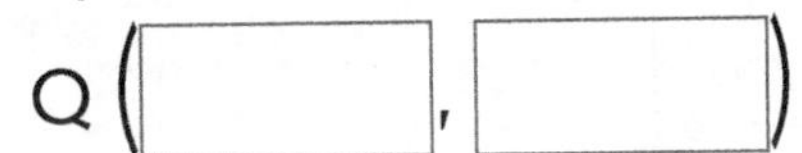

Q (,)

R (,)

S (,)

b) Join the coordinates to make a quadrilateral.

c) Name the quadrilateral. ____________

2. Plot and label a point, C, in the **4th quadrant** so that the points join to make an isosceles triangle.

C (,)

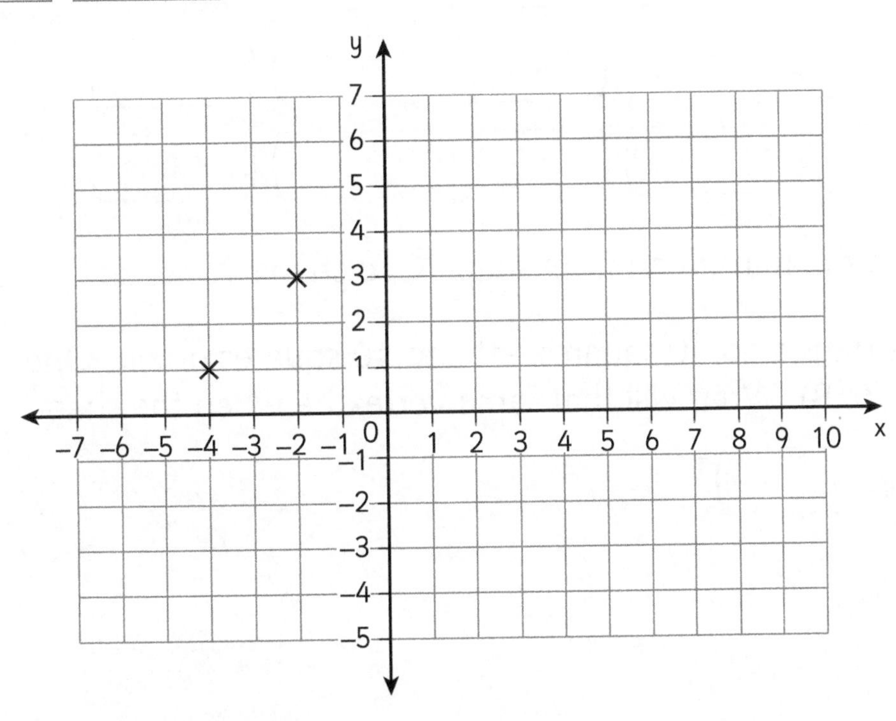

Unit 47: Draw and translate simple shapes on the coordinate plane, and reflect them in the axes

1.

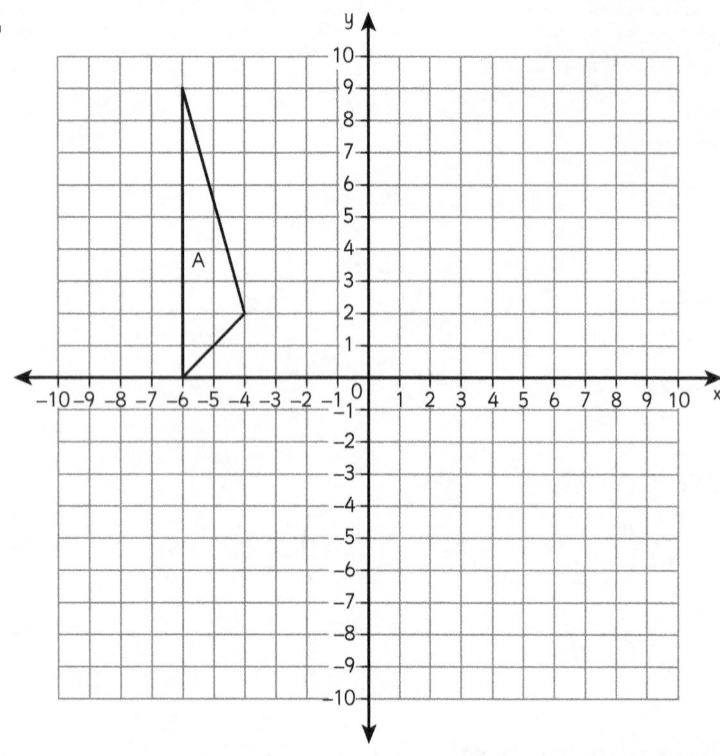

a) Translate shape A 10 squares to the right and 6 squares down. Label it B.

b) Translate shape A 4 squares to the left and 9 squares down. Label it C.

c) Write the sets of vertices for each shape.

A (☐ , ☐) (☐ , ☐) (☐ , ☐)

B (☐ , ☐) (☐ , ☐) (☐ , ☐)

C (☐ , ☐) (☐ , ☐) (☐ , ☐)

d) Describe the translation from shape C to shape B. ______________

2. A shape is translated 10 squares left and 10 squares down. One of the vertices is (6, 8). What will that same vertex be when the shape is translated?

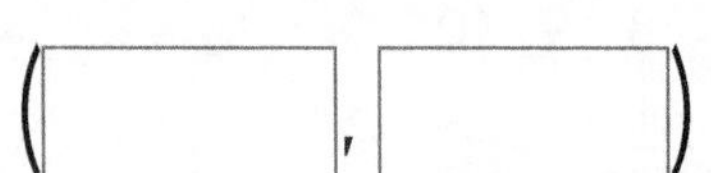

Unit 47: Draw and translate simple shapes on the coordinate plane, and reflect them in the axes

3.

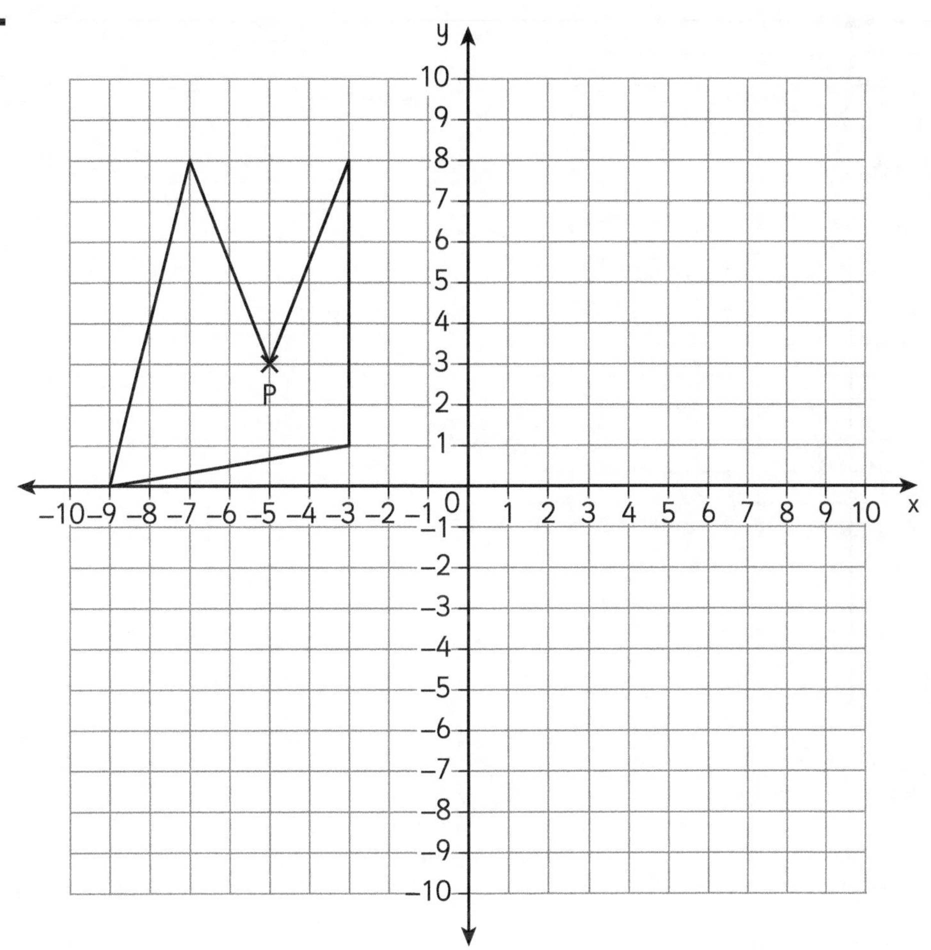

a) Reflect shape P in the *y*-axis. Label it Q.

b) Reflect shape P in the *x*-axis. Label it R.

c) Write the vertex marked with a cross in all three shapes P, Q and R.

P (___ , ___) Q (___ , ___) R (___ , ___)

Unit 47: Draw and translate simple shapes on the coordinate plane, and reflect them in the axes

1.

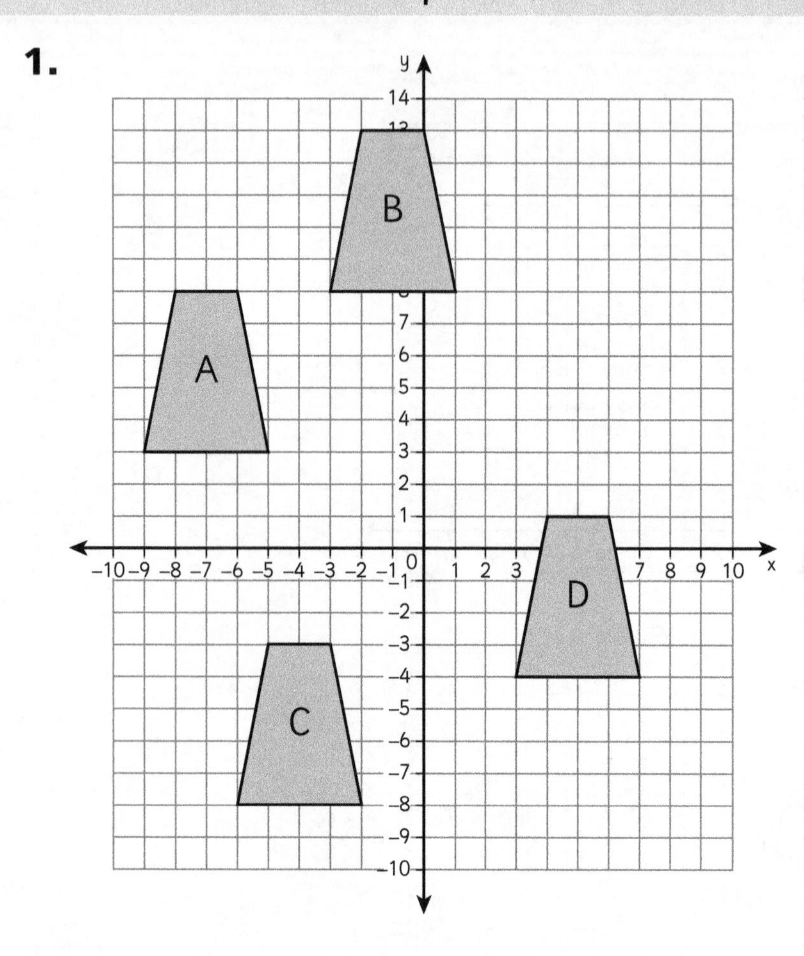

Describe the transformations from:

a) A to B ______________________________

b) C to D ______________________________

c) D to B ______________________________

2. A triangle with the vertices P (0, 4), Q (12, 6) and R (4, 0) is translated three squares to the left and five squares down. What are the vertices of the translated triangle?

P1 (____ , ____) Q1 (____ , ____) R1 (____ , ____)

Unit 47: Draw and translate simple shapes on the coordinate plane, and reflect them in the axes

3. Describe each transformation of A to B fully.

a)

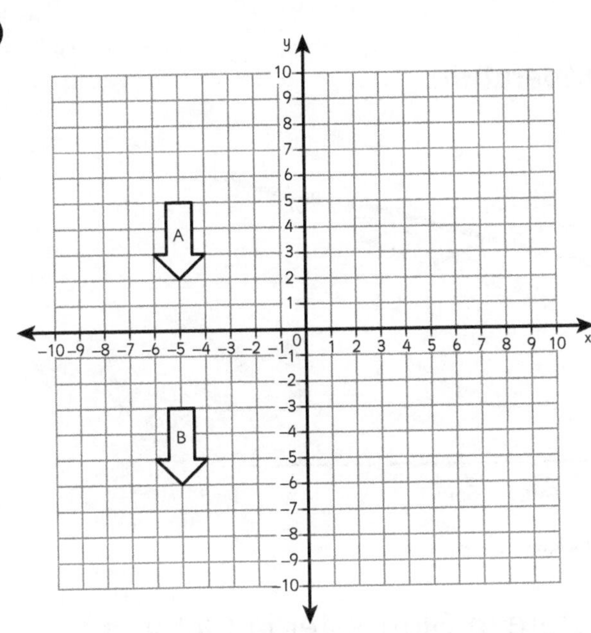

b)

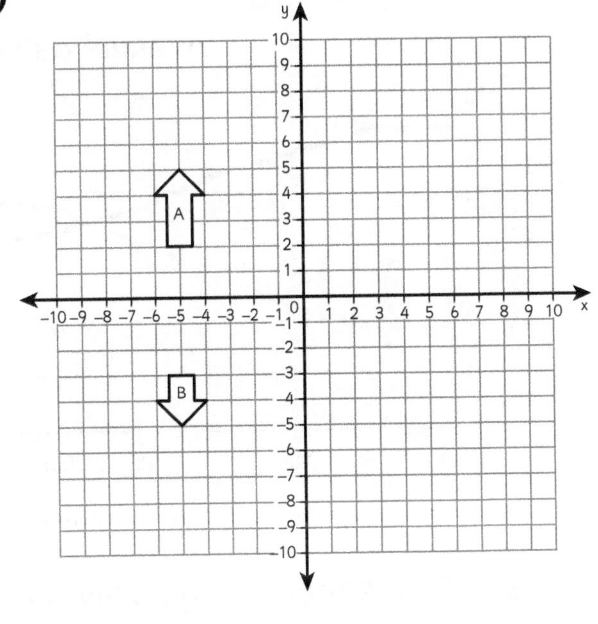

c)

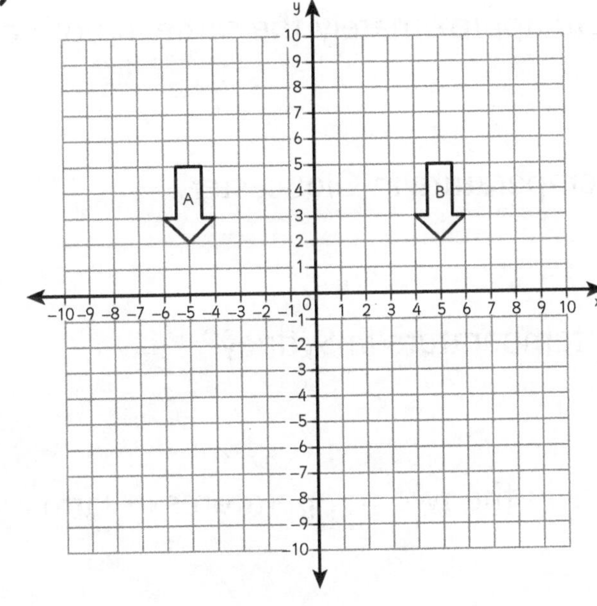

4 Complete the diagram to show a reflection in the *y*-axis.

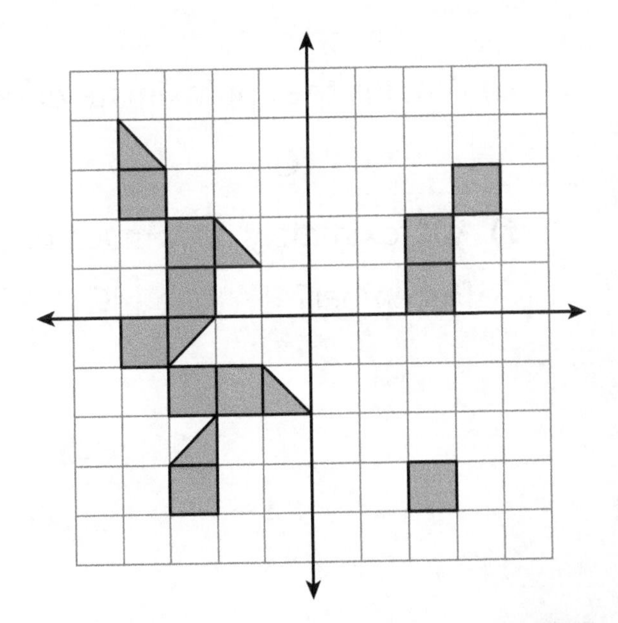

Unit 48: Interpret and construct pie charts and line graphs and use these to solve problems

1. The graph compares the average daily temperature of two cities: Gloucester, England and Sydney, Australia.

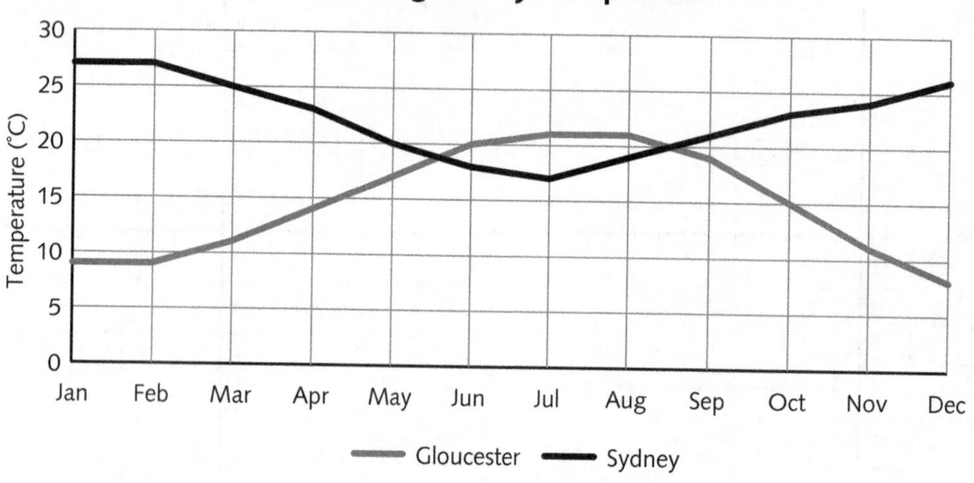

a) What is the average daily temperature in Gloucester in October?

[ ] °C

b) In which months is the temperature approximately the same in both cities?

____

c) Which month shows the lowest temperature in Gloucester?

____

d) Which months shows the highest temperature in Sydney?

____

e) What is the approximate difference in the two temperatures in January?

[ ] °C

f) Approximately how much hotter is it in Sydney than in Gloucester in December? [ ] °C

Unit 48: Interpret and construct pie charts and line graphs and use these to solve problems

2. The pie chart shows the proportion of the different food groups that make a balanced diet.

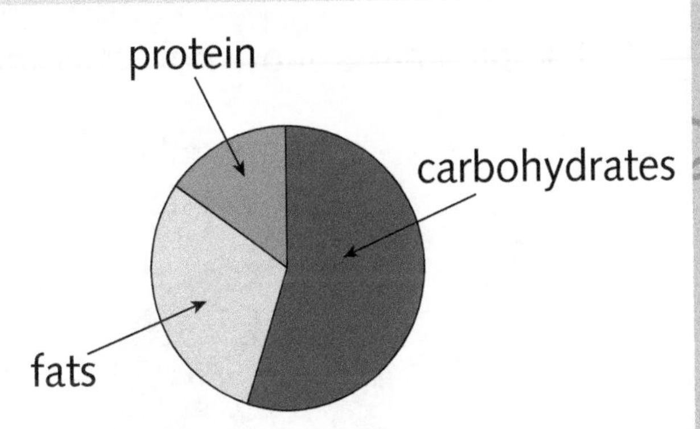

a) Write the approximate percentages for each food group. Make sure they add to 100%.

protein ☐ %   carbohydrate ☐ %

fats ☐ %

b) If a meal contains 100 g of protein, estimate how much carbohydrate is need to make a balanced diet.

☐ grams

3. Here are the results of a survey on Y6's favourite drinks. Use the information in the table to complete the pie chart. Work out the total, then think about how many people each sector represents.

Drink	Number
milk	8
water	11
hot chocolate	6
squash	10
fruit juice	5
Total	

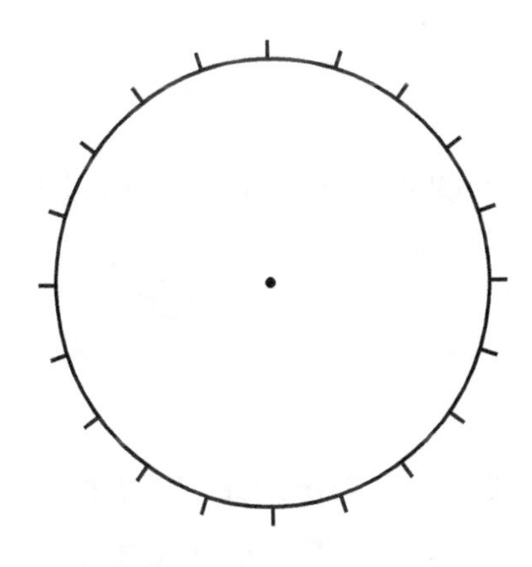

Unit 48: Interpret and construct pie charts and line graphs and use these to solve problems

1. The pie charts show how children in two different schools travel to school.

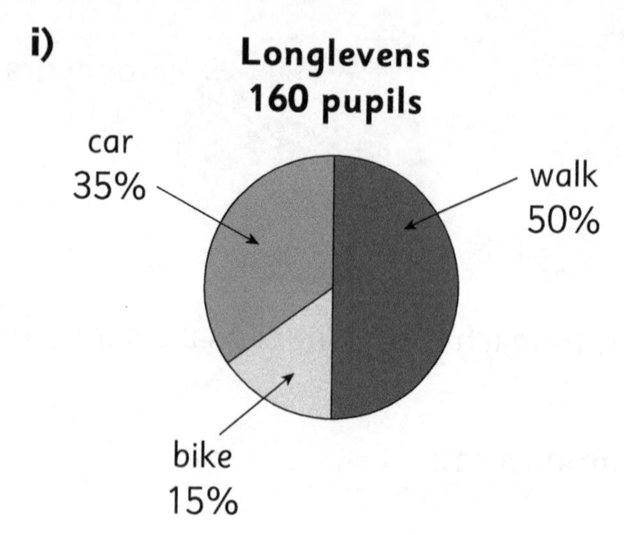

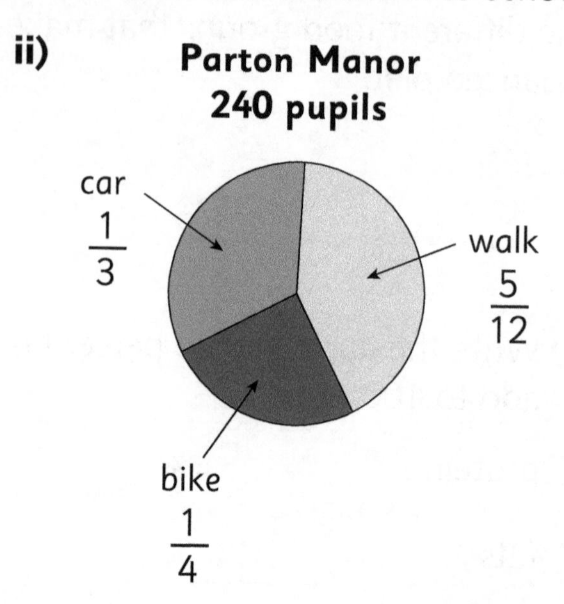

a) At which school do a **higher proportion** of children come to school by car? Show how you know.

____

b) At which school do more children walk to school? Show how you know.

____

c) How many more children ride a bike to Parton Manor than to Longlevens?

☐ children

2. Use a protractor to construct a pie chart with 8 equal sectors.

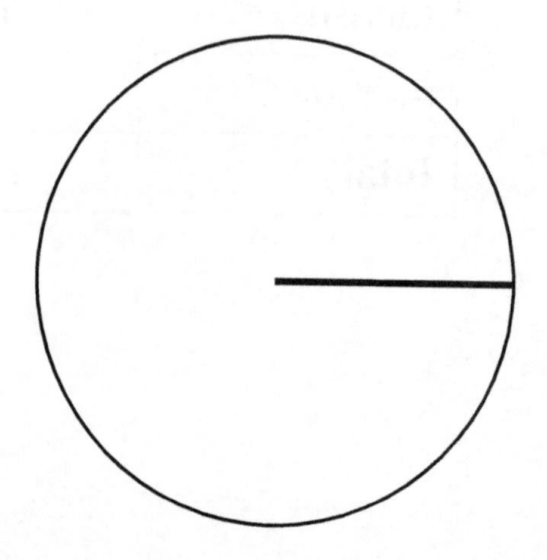

Unit 48: Interpret and construct pie charts and line graphs and use these to solve problems

3. This dual line graph compares the tariffs of two taxi companies.

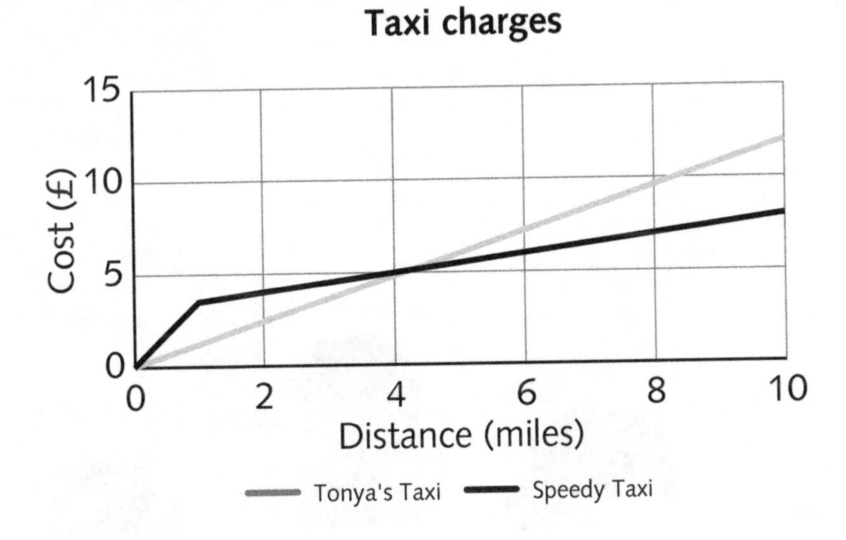

a) How much will a 5-mile taxi ride with Tonya's Taxi cost? £ ☐

b) Lukas paid Speedy Taxi £6.50. How many miles was his taxi ride?

☐ miles

c) When is it cheaper to travel with Speedy Taxi and when is it cheaper to travel with Tonya's Taxi? Explain your answer.

____

d) How much cheaper is an 8-mile taxi ride with Speedy Taxi, than with Tonya's Taxi?

£ ☐

e) Continue the graph lines to find the cost of a 15-mile taxi ride with both companies.

Speedy Taxi £ ☐ Tonya's Taxi £ ☐

Unit 49: Calculate and interpret the mean as an average

1. Find the mean of these sets of numbers.

a) 8, 4, 9, 3, 0, 6

mean: ______

b) 3·3, 4·1, 3·6, 5

mean: ______

2.

Ollie 1·2 m Chloe 1·15 m Zach 1·25 m Peyton 1·4 m Cohen 1·08 m

a) What is the mean height of these children?

______ m

b) Who is closest to their average height? ______

3. The table shows how many children are in each year group.
What is the mean (average) number of children per year group?

Year	Number
3	87
4	91
5	92
6	88

______ children

Unit 49: Calculate and interpret the mean as an average

1. Find the mean of these numbers.

a) 120, 135, 95, 125, 130, 115

mean: ☐

b) 1·5 kg 850 g 1·28 kg 1 kg 350 g

mean: ☐

2. Here are some number cards. The mean of the cards is 12. The mystery cards have a difference of 2. Write the numbers of the mystery cards.

3. Some marbles have been packed into bags. The mean number of marbles in each bag is 18. The table shows how many marbles each bag contains. 3 bags each contain 16 marbles and there are 5 bags left to fill equally with marbles. How many marbles should be put in each of the remaining bags?

Number of bags	Number of marbles
3	16
8	17
4	19
5	?

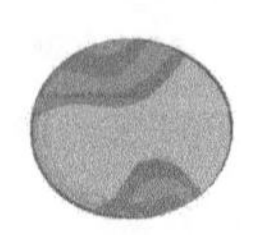
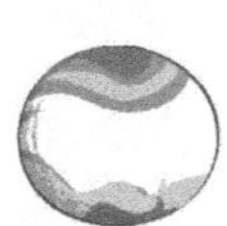
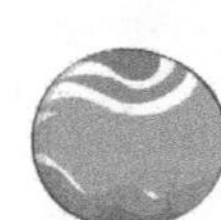

Notes

Notes

Notes

Notes

Notes